PHOTOGRAPHY THROUGH THE MICROSCOPE

Kodak books on photographic applications stem from early efforts of both George Eastman and, the first director of the Kodak Research Laboratories, C. E. K. Mees. While still at Wratten & Wainwright Limited in England, Mees authored or directed publication of many booklets for "the provision of technical information for those interested in photography." One of the booklets published by Wratten & Wainwright between 1906 and 1910 was on photomicrography. When Mees came to the United States to plan and direct the first Kodak Research Laboratory, he brought with him plans to continue technical photographic publications. In George Eastman, he found a willing supporter.

The first American edition of the booklet on photomicrography appeared with a 1919 copyright. Through the years, fourteen editions of the booklet have carried the title, *Photomicrography—An Introduction to Photography with the Microscope*.

In 1952, a new title was chosen when the photomicrography book was integrated into a series of Kodak Industrial Data Books. The book was now called *Photography Through the Microscope*. Succeeding editions have appeared, and now, with publication of a completely revised edition, this is the seventh to bear that title.

The current edition places emphasis on achieving optimum illumination in the microscope. As in all applications of photography, correct lighting of the subject is the key to superior photographic results. Once the lighting arrangement is achieved, camera positioning and exposure calibration fall easily into place.

For easy reference, film information and other data is located in appendices at the back of the book. To provide for easy location of other information, a comprehensive index is also included.

CONTENTS

John Gustav Delly

Much of this new edition of *Photography Through the Microscope* was written, revised, or edited by John Gustav Delly. John Delly is a senior research microscopist, scholar, lecturer on photography and microscopy, and author. John has also provided most of the new illustrations including the array of photomicrographs on the cover and the sequences on setting up Köhler illumination in the microscope. John's contributions have been edited and adapted by Kodak editors so that final responsibility for errors, inaccuracies, or omissions rests with those editors.

© **Eastman Kodak Company, 1980**
Seventh Edition
Standard Book Number 0-87985-248-8
Library of Congress Catalog Number 79-54858

Photography Through the Microscope

Photomicrography* is the technique of making photographs through a compound microscope. It involves coupling a camera to a microscope to produce enlarged photographs of very minute, microscopic detail. In fact, modern photomicroscopes incorporate the camera as an integral part of the microscope design.

The art and science of photomicrography are very old—as old as photography itself. Experiments were made in England in 1839 within months of the introduction of the Daguerreotype process from France, but sharp photographic images comparable to those produced by the objectives then in use were not successful until the perfection of the wet-collodion process in 1851. Still, photomicrography was well established in 1856, using wet-collodion negatives and silver-albumen paper.

Any type of microscope can be used to make photomicrographs, including hobby-quality or research-quality, brightfield or darkfield, inverted or erect, metallurgical, biomedical, polarizing, fluorescence, or any of the specialized interference microscopes. Because the principles apply to all types of microscopes, the discussion here centers on the so-called *brightfield microscope*. This type is commonly found in schools where biology is taught and in biomedical laboratories all over the world. In brightfield microscopy, the generally transparent or translucent specimen is either naturally colored or artificially colored (stained) and appears dark against a bright, white background or field; hence, the term brightfield. To be sure, other methods will be discussed, including the photomicrography of opaque specimens, such as metals, ores, and commercial and industrial products.

Cinemicrography is nothing more than a special form of photomicrography in which a motion-picture camera is used instead of a still camera. This type of photomicrography is necessary when recording dynamic events, such as live organisms, tissue cultures, and industrial processes. There are some important special considerations to be aware of when making films through the microscope and these are discussed in the appropriate section.

To make good photomicrographs, one must know how to use the microscope efficiently, and to use the microscope efficiently, one must have a good working knowledge of the instrument. This knowledge includes the selection, capabilities, and limitations of the optical components; how to adjust the microscope; how to illuminate the specimen; and even how to prepare specimens for examination and photography. Those who intend to make photomicrographs should also have some knowledge of photography, including an understanding of exposure and of sensitized materials and how a camera can be suitably attached to a microscope.

Photomicrographs are made for a variety of reasons. Teaching is one of the major fields that utilize photomicrographs. It is easier to project a color slide for class viewing than to have each member peer into the microscope and try to locate the structure under discussion. A photomicrograph is often used as a research record or to illustrate a particular phenomenon or condition in a published article. Photomicrographs of metals are very often made to supplement engineering or metallurgical reports.

Photomicrography is used in practically every field in which a microscope is the fundamental tool and whenever an enlarged, recorded image would prove useful. Photomicrography is important to teachers and students in high schools and colleges for laboratory studies in biology, botany, zoology, and anatomy. It is probably of even greater importance in industry, medical research, pathology, criminal investigation, agriculture, and forestry. Photomicrographs are even used in advertising. And, of course, photomicrographs may be made for purely aesthetic reasons.

Whatever the purpose of a photomicrograph, one must understand that the recorded image is no better than the image produced in the microscope, with the possible exception that contrast can be enhanced photographically. Thus, the first step in photomicrography is to understand the compound microscope and how to use it to best advantage in visual applications.

---

*Photomicrography should not be confused with *microphotography*, which involves making extremely small photographic images of large objects. The distinction between the terms photomicrography and microphotography was made as early as 1858, but the confusion persists. A contributing factor in the confusion is faulty translation from the German language in which photomicrography is *mikrophotographie*.

Chapter One

UNDERSTANDING THE COMPOUND MICROSCOPE

General Principles

A *simple magnifying system* uses a single lens unit to form an enlarged image of an object for viewing or for projecting. An example of a simple viewer is the common pocket magnifier or reading glass (Fig. 1-1). In slide projection, a transilluminated transparency is enlarged with a simple magnifying lens to form a real image on the screen.

In a *compound magnifying system,* magnification takes place in two stages. For example, in the projection situation above, if you remove the screen where the enlarged image is focused and instead locate a suitable lens behind this position, the lens would form another, greatly enlarged, image of the slide. The total magnification is the product of the magnification of the first lens and the second lens. If the second lens is used like a pocket magnifier, the final image will be a virtual one. This is the basic principle of a compound magnifying system—the one found in a compound microscope. The observer looks at the first, or primary, image with a lens that produces an enlarged secondary image, called a *virtual image.* That is the image the eye perceives. (See Fig. 1-2.)

Another feature of microscopes is that the lenses are of relatively short focal lengths. The shorter the focal length, the greater the magnification at a given image distance. In a microscope a high, two-stage enlargement is attained over a relatively short optical path with such short-focus lenses.

The first lens in a microscope is called the *objective* since it is near the object. This lens projects a magnified image of the specimen to a fixed position called the primary or *intermediate image plane.* The amount of magnification that the objective produces at this fixed distance is called its *magnifying power.* The magnifying power of an objective is classified as 1X, 5X, 10X, 20X—up to 100X and (rarely) higher. The primary image is located within, and about 1 cm down from the top of, the body tube of the microscope. The distance from the objective back focal plane to the primary image is called the *optical tube length.*

The second lens is placed in the body tube above the primary image. This lens is called the *eyepiece* and forms a secondary, further-enlarged image within the microscope. The

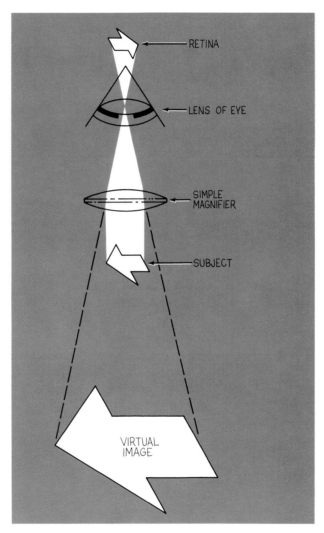

Fig. 1-1
SIMPLE MAGNIFIER—*A simple magnifier uses a single lens system to enlarge the object in one step.*

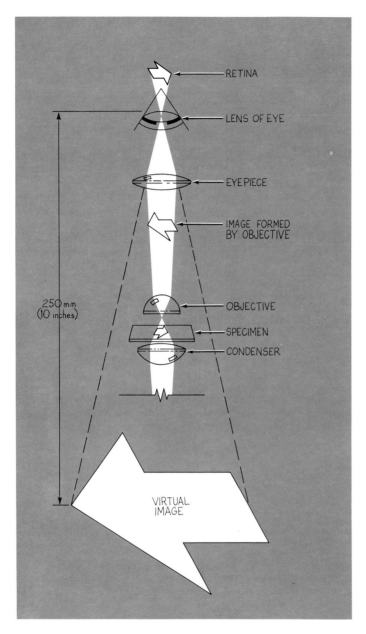

RETINA

LENS OF EYE

EYEPIECE

IMAGE FORMED
BY OBJECTIVE

250 mm
(10 inches)

OBJECTIVE

SPECIMEN

CONDENSER

VIRTUAL
IMAGE

Fig. 1-2
COMPOUND MAGNIFIER—
In the compound microscope, the intermediate image formed by the objective is enlarged by the eyepiece.

eyepiece (often called the *ocular*), like the objective, is classified in terms of magnifying power and could be 5X, 10X, or higher—up to 30X.

The total amount of image enlargement, or *magnification*, produced within the microscope is found by multiplying the magnifying power of the objective by that of the eyepiece. A 10X objective and a 10X eyepiece then produce a *visual magnification* of 100X, or X100 as it is sometimes written. This is the situation that is obtained when the objective is screwed to the bottom end of the body tube and the eyepiece is placed in the upper end of the body tube. Usually the length of the body tube is fixed—most commonly at 160 mm. Other *mechanical tube lengths* are 170 mm, 200 mm, or 210 mm. Some microscopes have body tubes of variable lengths in which one or two tubes telescope out from a larger body tube. The adjustable tubes are called *draw tubes* and are usually graduated in millimetres from about 145 mm fully closed to about 200 mm fully extended.

Modern large research microscopes that have provisions for illuminating apparatus above the objective usually have mechanical tube lengths considerably in excess of 160 mm. In this case, additional parallelizing lenses are placed within the system to shorten the apparent optical path so that the objective and eyepiece still see the total length, whatever it may be, as 160 mm. These lenses, however, introduce additional magnification within the system that must be taken in account when calculating total magnification. This additional magnification is referred to as the *tube factor* and is commonly 1.25X. Thus, the 10X objective and 10X eyepiece that produced a 10 x 10, or 100X, magnification in the above example will produce a 10 x 10 x 1.25, or 125X, magnification when used on a microscope with a tube factor of 1.25X.

As light rays emerge from the eyepiece they converge to a point called the *eyepoint*. This is the position that the eye seeks to see the whole image field when one looks into a microscope. The eyepoint is also often referred to as the *Ramsden disk* or *exit pupil*. In practical use, the lens of the eye is placed exactly at this spot. The distance from the eyepoint to the virtual image, or final image, within the microscope system is 250 mm (10 inches). Magnifying power of the microscope is arbitrarily based upon 250 mm, which is assumed to be the normal close viewing distance.

Instead of looking into the microscope, you can allow the image to be projected from the eyepiece onto a ground glass or photographic film. If the film plane or camera is 250 mm from the eyepoint, the magnification at the film plane will be the product of the eyepiece and objective magnifications. Longer or shorter distances will produce respectively higher or lower magnifications than the product of the eyepiece and objective magnifications.

The very fine details in a micro-specimen must be distinguished by the objective lens. This lens must be of high enough quality to resolve these details and produce an efficient primary image. The main purpose of the eyepiece is to further enlarge the image formed by the objective. This image can be degraded by an eyepiece of low quality, but it cannot be improved over what the objective lens presents to it (as far as image resolution is concerned), regardless of eyepiece quality.

Objectives

Microscope objectives are the most important components of the compound microscope. They form the *intermediate (primary) image* of the object, which is subsequently examined with the eyepiece. Microscope objectives are usually classified in terms of

- Magnifying power
- Numerical aperture (NA)—a measure of the light-gathering power and therefore, the resolving power
- Degree of optical correction—classified as achromat, semiapochromat (fluorite), or apochromat
- Mechanical tube length
- Cover glass thickness
- Flatness of field
- Focal length

Most, or all, of this information may appear engraved on the objective itself.

Fig. 1–3
MICROSCOPE OBJEC-TIVE—*The objective typically carries such information as the degree of correction, focal length, numerical aperture, and magnification.*

Correction of Aberrations in Objectives

	Chromatic	Spherical
Achromat	2 wavelengths	1 wavelength
Semiapochromat (Fluorite)	2 wavelengths	2 wavelengths
Apochromat	3 wavelengths	2 wavelengths

The *achromat* is the most common type of objective found on any microscope, accounting for 90 to 95 percent of all objectives in use. It is also the least expensive. In achromats, chromatic aberrations are corrected for two wavelengths (usually red and blue) and spherical aberrations are corrected for one wavelength (usually yellow-green, the wavelength to which the human eye is most sensitive). Achromats are very serviceable objectives. They have good working distance, moderate numerical aperture, and are adequate for most applications. If an achromat is used with white light, color fringes may appear in the outer margins of the image, because in this objective not all wavelengths are brought within an acceptable focus range. When black-and-white film is used to photograph such an image, these fringes may contribute toward a fuzzy image. If monochromatic light is used—that is, light of a single wavelength or a narrow band of wavelengths (such as green)—the image will be much sharper. Therefore, achromats produce the best image in black-and-white photomicrography when a green filter is placed in the light path. Images of inferior quality may result when light of a longer or shorter wavelength is used. The color fringing, especially at the edge of the field, will, of course, be recorded in color photomicrography when white light and achromats are used. A special variety of achromats is made for polarized light work. These are strain-free and are carefully made from the slow cooling of the glass to the final assembly of the lens system.

Fluorite objectives, or *semiapochromatic objectives* as they are sometimes called, are more highly corrected than achromats. Chromatic and spherical aberrations in these objectives are both corrected for two wavelengths. Note that this contributes an additional spherical aberration correction over the achromats. At a given magnification, the higher correction allows somewhat higher NAs than those afforded by achromats. Higher correction is achieved through the use of fluorite (calcium fluoride), which has a refractive index-dispersion relationship not obtainable in glass. Because natural, clean fluorite of optical quality is difficult to find in nature, it is expensive, increasing the cost of this type of objective. Recently, synthetic fluorite has been

produced and used to make microscope objectives. Fluorites are an excellent compromise between the more universal achromats and the more highly corrected apochromats, both in quality and in cost, and make excellent photomicrographic objectives. Fluorite objectives must be used with matched compensating eyepieces for finest image quality.

Apochromats represent the finest, most highly corrected type of objective available. Through the use of fluorite and special glasses, chromatic aberrations are corrected for three colors (red, green, and blue) and spherical aberrations, for two colors (red and blue). The NA of apochromats is higher than for others of the same magnification. This higher NA results in sharper and more crisp images compared to other objectives of the same magnification. Unfortunately, field curvature, or the inability to retain image focus all the way across the field of view at one focus setting, is increased; but this has been corrected in *flat-field objectives*. Generally speaking, the working distance, from the front lens mount to the top of the cover glass, is also decreased, but there are exceptions. Because of their high degree of correction for aberrations, the apochromats are particularly suitable for color photomicrography and for the best resolution of fine details and the finest image quality. Like the fluorites, apochromats must usually be used with matching compensating eyepieces. Recent designs may allow free interchange of eyepieces.

Field Curvature

Objectives exhibit a certain amount of *curvature of field* due to the different distances through which the on- and off-axis image-forming rays must pass. Depending on the setting of the focusing knob, the image can be made sharp in the center of the field, but with sharpness fall-off toward the periphery (see Fig. 1-4a); or the image can be made sharp at the periphery, but be out of focus at the center of the field of view. In visual microscopy, field curvature is not a problem because the microscopist always has one hand on the fine focus control and continually makes small adjustments while scanning the field. However, field curvature becomes a problem

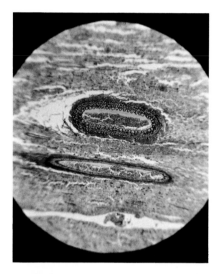

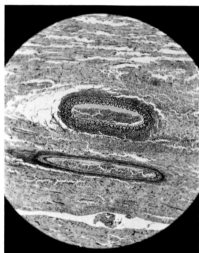

Fig. 1–4
FIELD CURVATURE—*The image at the top shows sharpness fall-off toward the periphery of the field due to field curvature. In the image below, field curvature has been corrected.*

in photomicrography. A microscopist is frequently judged a poor photomicrographer if the periphery of the photomicrograph is out of focus, as in Figure 1-4a, even though this is a property of the objective.

Field curvature can be overcome in several ways. One way to overcome this effect to some extent is to limit the recorded area to the center of the field when possible. This is done automatically with some attachment photomicrographic accessory units with built-in optical magnification factors of, say, ½X (*reduction* of 2X) where only the center of the field is used. Bellows extensions for 35 mm or larger format films can be increased until only the center portion of the field of view is recorded. One must be very careful here of introducing *empty magnification* (to be discussed later). Special eyepieces can also be used to reduce appreciably the curvature of the image.

The best way to overcome field curvature is to eliminate it completely through the use of *flat-field* objectives. Any of the three types of objectives described above can be had with flat-field. The prefix *plan* or *plano* is usually incorporated in the name. Thus, it is possible to have a *planachromat*, a *planofluorite*, or a *planapochromat*. The curvature of field has been corrected in the optical design of these objectives, and the corrective elements are usually built into each individual objective. There is one manufacturer who has incorporated a part of the correcting system within the nosepiece or body part of the microscope, necessitating the use of that manufacturer's stand with the objectives. Flat-field objectives are particularly suitable for examination or photomicrography of large fields. When used with eyepieces designated specifically for them, focus is uniform over the entire field of view from center to periphery. (See Fig. 1-4b.)

Cost may be a factor in selecting objectives for photomicrography, and it is interesting that the cost is about the same for a flat-field achromat and non-flat-field fluorite, or for a flat-field fluorite and a non-plano apochromat. To limit cost, the photomicrographer may decide between field flatness and higher correction. In general, a fluorite costs two to three times more than an achromat, and an apochromat costs five to six times more than an achromat. The increased cost is due to the additional corrective elements. As a matter of interest, the 100X achromat may contain six lens elements; the 100X planapochromat may contain as many as eighteen lens elements to effect all of its corrections. As in general photography, however, expensive equipment does not guarantee good images. Equipment is only one ingredient that contributes to the

final photomicrograph. For photomicrography in general, flat-field objectives are desirable but not essential.

Optical Aberrations

Aberration is a failure of a lens to produce exact point-to-point correspondence between an object and its image. Two optical aberrations mentioned in the discussion of objectives above are chromatic and spherical aberrations. If a simple, positive lens is used to project the primary, enlarged image, the quality of the image would be very poor due to inherent chromatic and spherical aberrations. In order to improve image quality it is necessary to design lenses so that aberrations are reduced.

Chromatic aberration occurs because the focal length of a simple lens varies noticeably with wavelength. Blue rays are shorter in wavelength and focus closer to the lens than green or red rays. (See Fig. 1-5a.) The single lens is unable to bring light of all colors to a common focus, resulting in a slightly different sized image for each wavelength at slightly different focal points. This situation prevailed in microscope objectives until about 1820. Correction for chromatic aberration originally applied to telescope objectives was then introduced into microscope objectives. By 1830 the achromatic objective was in fairly common use. Achromatism ("a" = without; "chroma" = color, i.e., images without color fringing) was achieved through the combination of two lenses of different optical properties cemented together to form a doublet. The proper selection of thickness, curvature, refractive index, and dispersion results in a lens that reduces chromatic aberration by bringing two wavelengths to a common focal point. (See Fig. 1-5b.)

A 10X achromatic objective typically contains two of these achromatic doublets. Triplets, lenses made with three cemented lenses, are also common in modern microscopy.

By 1886 Ernst Abbe perfected a complete series of apochromatic objectives, through the use of fluorite and new experimental glasses to make the doublets and triplets. Their great value for visual microscopy and photomicrography was instantly recognized.

Spherical aberration occurs when light rays passing through the central and outer portions of a lens are not brought to focus at the same distance from the lens. (See Fig. 1-6.) This condition arises in spherical lenses because light is refracted more at the edge of the lens, with gradual reduction to zero at the optical center. The image of a point is not reproduced

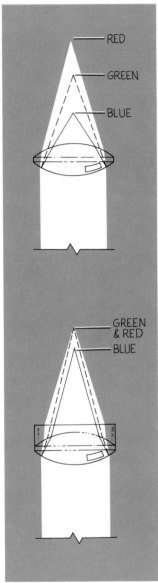

Fig. 1-5
CHROMATIC ABERRA-TION—*Failure of a simple lens to bring light of different wavelengths to a common focus (a) can be compensated in part by use of an achromatic lens (b).*

as a point but as a larger, circular area. The image of an object that is composed of an infinite number of points cannot possibly be sharp as long as spherical aberration is present. Also, white light is composed of all colors—which are refracted and focused differently, depending on specific wavelength. It is possible to correct spherical aberration for one color, but it might not be corrected for another color.

Stopping down a simple lens eliminates the worst spherical aberration from the edges, but optical correction yields better images. The sharpness of the image produced in a micro-scope is limited by the degree of correction for spherical aberration. As you shall see later, it is possible inadvertently to introduce spher-ical aberration in the final image even though you are using the most highly corrected objectives.

Numerical Aperture

In conventional photography, photographic lenses are always classified in terms of f-value, which is an indication of light-gathering power. A fast lens might be $f/1.4$ or $f/1.0$. Slower lenses might have f-values such as $f/4.5$, $f/5.6$, or $f/8$. The f-value of a photo-graphic lens is determined by dividing the focal length by the entrance pupil (apparent diameter of the diaphragm opening).

Microscope objectives, however, are not classified in terms of f-value although we are still interested in their light-gathering power, particularly their ability to capture highly diffracted image-forming rays from the spec-imen. It is these rays that allow fine structural details in a specimen to be distinguished. Magnification is required to bring the struc-tural details up to a size to be seen and recorded. The property of an objective that enables it to resolve fine detail is termed its universal aperture, or *numerical aperture (NA)* as it is usually called.

Mathematically, the numerical aperture is expressed as the product of the refractive index (n) for the medium in which the lens operates (for example, 1.00 for air, 1.51 for oil) and the sine of one-half the angular aper-ture of the lens (u):

$$NA = n \text{ sine } u$$

The angular aperture varies with the focal length of the objective. The angular aperture is the total maximum included angle of image-forming light rays from the specimen that the objective front lens can take in when the specimen is in focus. The idea of angular aperture is shown in Fig. 1-7.

Notice that as the focal length decreases, the maximum angle made between the spec-

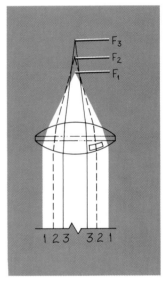

Fig. 1-6
SPHERICAL ABERRA-TION—*Failure of the lens system to image central and peripheral rays at the same focal point arises with spherical lenses. Optical correction is possible, but care must be taken not to introduce additional spher-ical aberration when setting up the microscope.*

Typical Objective Numerical Aperture (NA)

	10X	40X-50X
Achromat	0.25	0.65
Semiapochromat (Fluorite)	0.30	0.70-0.85
Apochromat	0.32	0.95 (dry)

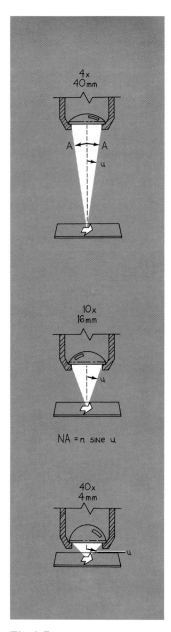

Fig. 1–7

NUMERICAL APERTURE— *Both light gathering power and resolving power are related to the angular aperture of the objective. Numerical aperture is expressed in terms of the angular aperture.*

imen and the diameter of the objective front lens is made to increase.* One-half of this angle, AA/2, is the term u in the above numerical aperture equation. Thus, the higher u is, the higher NA is. The importance of numerical aperture cannot be overemphasized; for, as you will see in the next section, the resolving power of the objective—the ability to distinguish between very closely spaced structural elements—depends largely upon the numerical aperture. The higher the NA, the higher the resolving power. Looking at the equation again, NA = n sine u, this means high n and high u.

What is the maximum possible NA? Theoretically, the highest angular aperture would be 180 degrees. One-half of that, u, would be 90 degrees, and the sine of 90 degrees is 1—which means that the NA is theoretically limited by n, the refractive index of the medium between objective and specimen. Most objectives are made to be used dry; air is the medium between the front of the lens and the specimen. Air has a refractive index of 1.00. The theoretical maximum NA possible for a dry objective is NA = n sine u = 1.00 × 1 = 1. In actual practice, the highest numerical aperture for a dry objective is about 0.95. The numerical aperture for any objective is engraved on the objective itself, usually under or next to the magnification.

NA depends, to some extent, on the degree of correction to make higher angular apertures possible. For example, a 10X achromat commonly has an NA of 0.25. A 10X fluorite objective commonly has an NA of 0.30. And a 10X apochromat has an NA of 0.32. Likewise, most 40 or 50X achromats have NAs of

0.65; fluorites of the same magnification have NAs of 0.7 to 0.85, or even 0.9; and dry apochromats in the same magnification range have NAs of 0.95.

Looking at the numerical aperture equation again, NA = n sine u, note that the only way to have an objective of numerical aperture equal to or greater than 1.00, is to place a liquid medium of higher refractive index (n) between the lens and the specimen slide. An *immersion lens* is designed for this purpose. Dry lenses should *never* be used with any liquid immersion media. If in doubt, look at the NA on the objective. If the number is less than 1.00, *do not immerse* the front lens in a liquid.† Objectives intended for immersion will almost always indicate the immersion medium. Several common media include water (n = 1.33), glycerin (n = 1.47), and immersion oil (n = 1.515). These should not be interchanged; that is, an objective designed for water immersion should not be used with oil. If you run out of immersion oil, do not use glycerin. Each immersion objective must be used with the medium it has been designed for. Recently, immersion objectives have been designed for use with any of the three common immersion media. The highest theoretical NA using common immersion oil is 1.5. In practice, NA only approaches 1.4, and 1.3 or even 1.2 is more common.

Most objectives in the 40 to 63X range are also designed for and must be used with immersion media. Their NA's are typically 1.00 but can go up to 1.40. The practical use of the immersion objective is described in a later section.

Range of Objective Numerical Aperture (NA)

Magnification	Numerical Aperture (NA)
4X	0.032-0.12
10X	0.17 -0.32
16X-25X	0.32 -0.65
40X-55X	0.57 -0.95 (for dry objectives)

---

*As the focal length decreases, magnification increases and the whole lens becomes smaller. The NA does not *necessarily* increase except as made to do so for resolution or brightness.

†A possible exception is when the immersion principle is applied to low-power, low-NA water immersion objectives used by biologists.

Resolving Power

The capacity of the optical system in a compound microscope to distinguish and separate fine structural details in a specimen is known as *resolving power*. However, this value is subjective. An image may be unsharp but may still be considered resolved. Resolving power is limited by the NA of the objective, but it also depends upon the working NA of the substage condenser. The higher the effective NA of the system (limited by the NAs of individual components, see previous page), the greater will be the resolving power. Resolving power is also dependent on the wavelength of light. The shorter the wavelength, the better the resolving power. Resolving power refers to the ability to distinguish two closely spaced structural elements. High resolving power means being able to detect ever smaller and smaller spaces between two things. (In photography, resolving power is usually referred to in terms of line pairs per millimeter. This is the reciprocal of the number found when computing R in the equations below.)

There are many equations for computing resolving power, at least a half dozen of which take into consideration both theoretical factors and psychophysical reactions of the observer. All have valuable, particular applications, but one that lends itself to the present discussion is the commonly seen

$$R = \frac{\lambda}{2NA}$$

where R is the smallest distance between two structural elements, λ is the wavelength of the light used, and NA is the numerical aperture. When the optical system is correctly aligned and adjusted, the NA of the objective can be used in the equation. In general, you want R to be small—the smaller the better. To have R small, the numerator should be small (short wavelength—toward the blue or violet), and the denominator should be large (high numerical aperture). The same units must be used for R and λ.*

Highest resolving power is obtained with ultraviolet radiation, which represents the shortest usable wavelengths. In the visible region of the spectrum, blue light has the next shortest wavelength, then green, and then red. If white light is used, the applicable wavelength is that for green—the middle of the visible spectrum and region of highest visual acuity.

---

*Other equations for resolving power according to Lord Rayleigh are

$$R = \frac{0.61\lambda}{NA} \quad \text{or} \quad R = \frac{1.22\lambda}{NA_{objective} + NA_{condenser}}$$

If the dominant wavelength for green light (0.550 micrometre) and an achromat of high NA (1.25) are applied in the formula, the resultant resolving power is 0.22 micrometre. With green light, then, it is possible to resolve structures down to 0.22 micrometre (0.00022 millimetre). The following table shows the effective resolving power of several achromatic objectives of different numerical apertures when green light is used.

Resolving Power for Achromats (green light)

Magnifying Power	4X	10X	20X	45X	100X (oil)
Numerical Aperture	0.10	0.25	0.50	0.85	1.25
Resolution (micrometres)	2.75	1.10	0.55	0.32	0.22

Resolving Power for Apochromats (green light)

Magnifying Power	4X	10X	25X	40X	100X (oil)
Numerical Aperture	0.16	0.32	0.65	0.95	1.40
Resolution (micrometres)	1.72	0.86	0.42	0.29	0.20

Change in Resolution with Wavelength

	Green	Blue	Ultraviolet
Wavelength (micrometres)	0.546	0.436	0.365
Resolution (micrometres)	0.195	0.156	0.130

The resolving powers for apochromatic objectives are better than those for comparable achromats, since they are more highly corrected optically and have higher numerical apertures. The above table shows pertinent, comparable data for apochromats, again with green light.

Resolving power for an apochromat of the highest NA (1.40) with light of different wavelengths is shown in the last table. This indicates the improvement of resolution with shorter wavelengths of light. It also indicates that we cannot expect to do better than about 0.2 μm visually and 0.13 μm photographically using 365 nm ultraviolet. Furthermore, these examples assume perfect optics, alignment, and observer.

Although it cannot be expected that anyone will often wish to determine the exact resolving power for any objective, it is important to understand the capabilities and limitations of a lens in order to use it to best advantage. This knowledge should help in the selection of a lens or of the type of light needed to photograph a specimen. What has been discussed here is the achievement of the

highest resolving power by best use of the optical components of the microscope. It must be noted that the contrast of the photographic material used for photomicrography can also influence the resolving power of the system.

Diffraction Theory and Resolution

The whole basis of numerical aperture and resolving power goes back to the specimen and diffraction theory. The theory of microscopical image formation, formulated over a hundred years ago, is based on the fact that light is diffracted when it passes through narrow slits or other structural elements. When light passes through parallel slits, for example, a primary or zero-order diffraction ray passes undeviated straight through, while to either side of the zero-order ray there will be first-order diffraction rays at some angle from the zero order that depends on the spacing between the slits. Beyond the first-order diffraction, at some greater angle, will be second-order diffraction rays, and so on. The important observation is that as the distance between the diffracting structures becomes smaller, the angles of diffraction become greater.

Now, only if all the diffracted rays enter the objective can interference take place to recreate the image in the intermediate image plane. The zero-order by itself is insufficient to produce a recognizable image of the specimen. If the zero-order and one of the first-order diffracted rays enter the objective, some semblance of the structure will be produced, but only if all of the rays are recombined will the image represent the true structure. It can readily be understood now why objectives of small angular aperture (see Figure 1-7) will not reveal the finer structures of specimens. The fine structure causes high angles of diffraction that never enter the objective. Conversely, it also explains why objectives of high angular aperture, i.e., high numerical aperture, are necessary to capture highly diffracted rays.

This point is of the utmost importance in photomicrography because the angular aperture, and therefore numerical aperture and resolution, can be controlled through the use of the aperture diaphragm in the condenser. Use this diaphragm indiscriminately and you may not get the resolving power in your photomicrographs that the objective is capable of yielding. The proper use of the aperture diaphragm will be discussed in the section on setting up correct illumination.

It should be noted that in actual fact, diffracted rays are always coming from the specimen at all angles. Since the objective is above the specimen, a high-NA objective will capture diffracted rays that are at an angle greater than the condenser illuminating cone. Evidence for this can be seen at the objective back focal plane where the haziness over the aperture diaphragm image is due to the more highly diffracted rays. These are strictly cut out not by the aperture diaphragm alone, but by a diaphragm in the objective—if indeed one would want to eliminate them. This is the basis for specifying the NAs of both the objective and the condenser in many resolving-power equations.

Working Distance of Objectives

The distance between the front lens of an objective and the top surface of the cover glass on a specimen slide is called the *working distance* for the objective. Working distance governs the allowable movement of the objective in obtaining critical focus of the specimen image. In the general case, working distance decreases rapidly as the focal length of the objective decreases and magnification increases. For oil-immersion objectives, the working distance is measured in fractional parts of a millimetre.

Cover-glass thickness is of relatively little importance optically with oil-immersion objectives since the oil has about the same refractive index as the cover glass.* The allowable movement of an immersion objective, however, is affected by cover-glass thickness. More working distance is obtained when the cover glass is thinner than specified or when no cover glass is used, as is often the case with blood smears. A thicker cover glass will decrease the working distance; the thickness can then approach the point where focus of the specimen may become impossible. Other minor factors that can affect the allowable working distance for an immersion objective are the viscosity of the oil and the amount medium between the cover glass and the specimen. The effects of these factors are small, but when they are added to the cover-glass thickness, they contribute toward a reduction in working distance.

The following table shows the average working distances for different objectives of various focal lengths. The figures can only be considered approximate since objectives of different manufacture will differ in characteristics. They are, however, fairly typical.

Objective Type	Magnification	NA	Focal Length (mm)	Working Distance (mm)
Achromat	10X	0.25	16	7.70
Apochromat	10X	0.30	16	4.85
Achromat	20X	0.50	8	1.60
Apochromat	20X	0.65	8.3	0.50
Achromat	45X	0.85	4	0.30
Apochromat	47.5X	0.95	4	0.18
Achromat	97X oil	1.25	1.8	0.13
Apochromat	90X oil	1.30	2.0	0.12

Working distance is a very important characteristic of objectives for microscopists who must manipulate or perform operations on their specimens, as, for example, biomedical microscopists working with tissue cultures or industrial microscopists performing microchemical tests or testing microelectronic circuits or working with the hot stage. For these and for the general photomicrographer who must take photomicrographs through deep vessels, special so-called *long-working-distance* objectives are made.

Tube Length

The objective is screwed† into the bottom part of the microscope body tube, while the eyepiece is inserted into the top of the microscope body-tube. The distance between the insertion position for the objective and the top of the draw-tube is called the *mechanical tube length.*

Most manufacturers specify a mechanical tube length of 160 mm. (Some previously specified 170 mm, but now 160 mm is generally considered universal for standard microscopes.) Objectives are designed optically for a specific tube length. If they are used at a different tube length, they will not retain their intended optical efficiency.‡ They should not, therefore, be interchanged on different microscopes if the specifications are not known. Image quality may suffer if this practice is followed. Some microscopes are equipped with adjustable drawtubes, so that mechanical tube length can be altered to comply with the specification for an objective. Microscopes with adjustable drawtubes were common in the past. They are, unfortunately, rare today—but perhaps with good reason.

Given a perfectly prepared specimen, poor image quality will result if the correct tube length is not used. Too great a tube length will result in overcorrection, and too short a tube length will result in undercorrection; either will result in poor definition (image sharpness). The reason for this is that changing the *mechanical* tube length for which the objective was designed will also change the *optical* tube length. The optical tube length is the distance between the objective back focal plane and the intermediate image plane. Changing the optical tube length introduces spherical aberration into the system. Dry objectives of high numerical aperture are particularly sensitive to changes in optical tube length.

One way to correct the poor image quality due to too thick or too thin cover glass is to deliberately introduce spherical aberration of

---

*Cover-glass thickness *is* important *in air.* See **Selection of Cover Glass,** page 16.

†The standard adopted by the Royal Microscopical Society (RMS) for screw thread of objectives seems to be the only one to which microscope manufacturers consistently adhere.

‡Tube length is not fixed for *infinity-corrected objectives.* Instead, such objectives require internal relay lenses in the microscope. These objectives should only be used with a microscope matched to their specifications. The objectives cannot be used on a conventional microscope with any tube length.

opposite effect to counteract and cancel the aberration due to the specimen preparation. The introduction of correcting aberration is made by changing the tube length. The draw-tube is shortened for too-thick cover glasses, and lengthened for too-thin cover glasses. When making a photomicrograph with a high-dry objective, for instance, the specimen is first brought into focus and then the draw-tube length is changed slightly at random, perhaps lengthened a bit. Changing the tube length makes it necessary to refocus the image.

After refocusing slightly, the microscopist must judge the image and compare it to that before the change. If the image is sharper, the microscopist should continue to increase the tube length in small increments. Normal practice is to proceed until the image deteriorates and then work backward again, locating the drawtube position for optimum image quality. It is at this tube length, whatever it may be, that the photomicrograph should be made. Of course one may have to sacrifice some image quality for this convenience.

When an objective is used at other than its specified tube length, its magnifying power will be different from that indicated. For example, a 10X objective designed for 160 mm tube length will magnify more than 10X if used at a 170 mm or longer tube length. Conversely, magnifying power will be reduced at lengths shorter than the intended tube length. A 10 mm difference in tube length amounts to a magnification difference of only 6 percent. This change of magnification with mechanical tube length could be taken advantage of, particularly when calibrating an eyepiece micrometer. The tube length could be changed to change the magnification slightly so that the division in the eyepiece micrometer would exactly equal some convenient unit of length, making size of particles and measurement of structures much easier.

Eyepieces

Eyepieces, or oculars, are exceedingly important in producing good visual and photographic images. The eyepiece in the compound microscope serves as the second stage of magnification, enlarging the primary image formed by the objective in the intermediate image plane. The eyepiece renders this real primary image visible as a *virtual* image when the microscope is used visually or projects it as a *real* image that can be recorded in a camera. Since the eyepiece acts as a magnifier to examine the real image formed by the

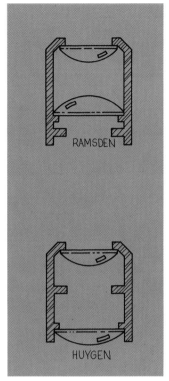

Fig. 1–8
EYEPIECES

objective, its optical quality and correction are very important.

There are several types of eyepieces ranging from the Huygenian, or negative, eyepiece used with most achromatic objectives to the more complex positive types used with high-power achromats, fluorites, and all apochromats. Simple eyepieces have only two lenses—a *field lens* closer to the objective image and an *eye lens* closer to the eye. These two lenses are planoconvex. In the Huygenian eyepiece, the plane side is uppermost in both lenses, and the eyepiece diaphragm lies between them inside the eyepiece. In the positive, Ramsden eyepiece, the plane sides are outermost and the diaphragm lies outside and beneath the two lenses.

The eyepiece diaphragm defines the round field of view one sees when looking in a microscope. This is also where graticules and reticles are placed, for the plane of the eyepiece diaphragm is where the primary image from the objective lies. Whatever is placed on the eyepiece diaphragm (measuring scale, cross hair, pointer), will be superimposed on and in focus with the specimen image.

Simple eyepieces are suitable for most achromats, but more highly corrected eyepieces are necessary for better objectives. Highly corrected eyepieces have a doublet for an eye lens and go by different names with different manufacturers—for example, Periplan; some eyepieces have a triplet here. Whenever high-quality objectives are to be used for photomicrography, the eyepiece should be of the *compensating* type to provide the high image quality and flat field of which the objective is capable. These compensating eyepieces, using doublets and triplets, are important because they correct residual aberrations in the objective image. Apochromats inherently produce what is called *chromatic difference of magnification.* Since the methods of correcting for these aberrations vary with manufacturers, it is essential to use only those eyepieces recommended by the manufacturer of the objective. Also, the position of the image produced by objectives of different manufacture varies, e.g., 10 mm, 13 mm, etc, from the top of the body tube. Since this position is also the location of the diaphragm in the eyepiece, use of eyepieces of a different manufacturer will not be parfocal with all of the objectives and may introduce field curvature.

Choosing eyepieces, then, is relatively simple. Pick the objective first; then buy the eyepieces made to be used with that objective. Incidentally, simple eyepieces will appear to have a blue ring all around the edge of the diaphragm (the field of view) when viewed while holding the eyepiece up to a light or on the microscope. Compensating eyepieces will show a red-orange color at the edge of the eyepiece diaphragm.

Visually, wide-field eyepieces produce a pleasing subjective impression of the field, and they are good for lengthy viewing sessions. For scanning smears and films and making counts, they save valuable time because they encompass such a large area. Wide-field eyepieces are not absolutely necessary for photomicrography, but there are one or two instances where they are beneficial, particularly when using certain homemade photomicrographic cameras and when using cameras with fixed lenses.

High-eyepoint eyepieces are made for those who must keep their eyeglasses on when viewing through the microscope. Eyeglass viewing is necessary when eye defects cannot be corrected with the microscope itself, e.g., astigmatism. People who are near-sighted or far-sighted may use the microscope without their eyeglasses if they wish. High-eyepoint eyepieces are even popular with those who do not wear eyeglasses. The high clearance (up to 25 mm from top of lens to eyepoint) reduces fatigue in prolonged viewing. There are also good reasons for using high-eyepoint eyepieces in photomicrography. With commercial photomicrographic cameras having long intermediate bodies, it would be otherwise difficult to get the eyepoint high enough to be in the plane of the shutter or lens. These eyepieces are also useful with ordinary cameras having fixed lenses. Eyepieces which combine high-eyepoint and wide field are very popular and very useful in photomicrography.

When any counting, measuring, or other procedure involving a graticule or reticle is to be undertaken, either visually or photomicrographically, the eyepiece should have a *focusing eye lens.*

Eyepieces for use with polarizing microscopes are keyed so that they can fit only one of two positions in the slotted body tube. This insures that the cross hairs will correctly indicate the vibration directions of the polarizer and analyzer. It is a good idea to buy a graticule that combines cross hairs and an eyepiece micrometer on the same disc to avoid the expense of an extra eyepiece.

Some eyepieces are intended specifically for photomicrography. These are frequently negative eyepieces which cannot be used visually at all. They give somewhat flatter fields with non-flat-field objectives. Still others have focusing eye lenses that may be adjusted for projection distance (distance from eyepoint to film) and are called, appropriately enough, *photo eyepieces.* Besides producing flat-field results, they are also color-corrected and therefore of advantage in color photomicrography. These will be discussed again in the section on cameras.

Eyepieces are produced with different magnifying powers ranging from 1X to 30X or more. Those in the 1X, 3X, and 5X range are usually confined to projection microscopes, and those of 25X to 30X are for special applications. Eyepieces of 5X to 20X are most commonly used in visual work with the 10X, 12.5X, and 16X the most popular and useful. In photomicrography, a wide range of eyepieces 5X to 25X, is used.

Condensers

The third major optical component of a compound microscope is the condenser, also called the *substage condenser.* The condenser provides a cone of light that illuminates the specimen. It is probably the most neglected, least understood, and most misused component of the microscope. Anyone now using a microscope who does not know the numerical aperture (NA) of the condenser (or who does not know what it means even after looking at it) probably does not understand the condenser's correct adjustment and use.

Light from the condenser converges on the specimen in the plane of the stage, and the light diverges in passing through the specimen to form an inverted cone whose included angle fills the objective front lens. The angular size of the illuminating cone is controlled by a variable diaphragm located beneath or within the condenser. This diaphragm is called the *aperture diaphragm;* in transmitted-light microscopes, it is also called the *substage diaphragm. Aperture diaphragm* is probably a more accurate term because it not only defines its purpose, controlling the size of the illuminating aperture, but it can also be applied to both transmitted-light and reflected-light microscopes. The correct focus of the condenser and the proper opening of the aperture diaphragm are of extreme importance in both visual microscopy and photomicrography. They should be adjusted to achieve Köhler illumination. (See page 28.)

Four principal types of condensers are available: (a) Abbe, (b) aplanatic, (c) achromatic, and (d) aplanatic-achromatic. The Abbe condenser is the simplest and least expensive type. It is provided on all microscopes unless another type is specified. It may have two or, less likely, three lens elements, and it is *uncorrected* for spherical or chromatic aberrations. One of the most important points to consider in choosing a condenser is the numerical aperture. The condenser NA should be equal to, or greater than, the highest objective NA—usually about 1.25 to 1.32 for the 100X objective. The effective NA of the system, and hence the resolving power, is limited by the lowest value for individual components of the system.

The refractive index of air is 1.00. This is significant here for two reasons. First, if the condenser NA is less than 1.00 (0.95, for example), it is not intended for immersion. It may be an excellent dry condenser, but the resolving power of a 1.32 NA objective cannot be realized with this condenser. The effective NA of the system can only be 0.95 even if the substage diaphragm is wide open,

which is seldom the case. Second, a condenser with an NA over 1.0, say 1.3, is intended for immersion. If the top lens of the condenser is not immersed in the appropriate liquid on the bottom of the slide, the effective NA will be that of the resulting air space, 1.00. To get full resolving power from objectives of NA greater than 1.0, one must use a condenser of NA greater than 1.0, matching or exceeding the objective NA, and the top lens of the condenser must be immersed in liquid on the bottom of the slide. Some condensers are provided with interchangeable top lenses which change the NA. Still others have a lever-operated, flip-out top lens or slide-in lower lens to provide at least two maximum NAs.

Besides the NA of the condenser, the degree of correction must be considered. An ordinary two-lens condenser will form an image of the field diaphragm in the field of view that will be somewhat fuzzy and surrounded with color fringes. This is caused by spherical and chromatic aberrations in such condensers. Condensers that focus light in one plane are termed *aplanatic.* There may be three, four, or even five lenses in such a condenser. In *achromatic-aplanatic* condensers, chromatic aberrations are also corrected. Such condensers may have five, six, seven, or more elements and are essential for truly critical microscopy. A two- or three-lens condenser is perfectly adequate for a student microscope or for one used only occasionally for noncritical, routine procedures.

Good photomicrographs can be made with any kind of condenser if it is used properly, but the best color photomicrograph requires achromatic or aplanatic-achromatic condensers. The image of the field (lamp) diaphragm in the specimen plane will be sharp, of good contrast, and free of color fringing when focused with an achromatic condenser.

Condensers for special purposes include strain-free condensers for use with polarized light. Darkfield, interference-transmission, interference-contrast, phase-contrast, and long-working-distance condensers are all special variations that are fully described in standard texts and some specialized papers. Quartz condensers and condensers built according to reflection principles are made for use with ultraviolet radiation.

An important factor in the optical system associated with condensers, but almost never considered, is *slide thickness.* Thickness of the microscope slide is just as important in relation to the condenser as the cover-glass thickness is to the objective. Condensers of high numerical aperture have shorter working

Correction of Condensers

Type	Aberrations Corrected	
	Spherical	Chromatic
Abbe	–	–
Aplanatic	X	–
Achromatic	–	X
Aplanatic-achromatic	X	X

distances than those of low NA. If the microscope slide is too thick, it will be impossible for a high-NA condenser to form an image of the field (lamp) diaphragm in the plane of the specimen. At best it will be located somewhere in the slide below the specimen. Slides whose thickness is not specified should not be purchased for critical microscopy and photomicrography.

How thick should microscope slides be? The answer to this question is not routinely supplied by microscope manufacturers, but slides of 1 mm thickness should work with all condensers, and a range of about 0.96 to 1.06 mm is ideal. Some private-label slides are as much as 1.25 mm thick. Except for certain darkfield condensers which specify this thickness, 1.25 mm is too thick for high-NA achromatic condensers. With a separate illuminator, the working distance of the substage condenser is affected by the optical distance of the lamp condenser.

Other Substage Components

Besides the condenser there are several components beneath the stage of a transmitted-light microscope, including a condenser centering device, mirror or built-in light source, filters, polarizers, and other special apparatus.

For critical photomicrography the substage should be provided with condenser centering screws so that the condenser itself can be correctly aligned on optical axis with respect to the objective.

A mirror is provided as part of the substage when a built-in light source is not included. Generally, the mirror is about 50 mm in diameter and is usually two-sided, with one side a plane mirror and the other, concave. The mirrors are usually second-surface mirrors that will form at least three images of the field diaphragm. The plane side of the mirror is always used with a condenser. The concave side of the mirror is used for very low-NA objectives when no condenser is used. A mirror is a relatively inexpensive item, but an important one.

The critical microscopist will, of course, want to buy a first-surface mirror. This may be of polished stainless steel or evaporated

aluminum protected with a thin silicon monoxide layer. This feature is especially important for photomicrography because a second-surface mirror will always have some component of the outside two images of the field diaphragm (one from reflection from the top surface of the glass, the second from second internal reflection) in the field of view. These may not even be seen visually but will be recorded on the film. Inexpensive first-surface mirrors are available in 50 mm discs from many optical supply houses. One of these can be quickly attached to an existing second-surface mirror with double-sided adhesive tape to eliminate this source of poor photomicrographs.

Built-in light sources are discussed in the section on Köhler illumination.

Polarizing microscopes have a graduated, rotating polarizer in the substage position. If polars are purchased for qualitative polarized light work and for photomicrography, they should be selected for color. Those nearest to absolute neutral gray, and not green, brown, blue, or magenta, will serve best. Except where high heat is encountered, wavelengths outside the visible are used, or absolute extinction is required, plastic polarizing material is acceptable. For the exceptions noted, calcite prisms are necessary.

Stages

Stages on which the microscope slide preparation is placed for examination may be square or circular and fixed, rotating, or interchangeable. Most stages are square and fixed despite the fact that the circular, rotating stage is far more useful, both for orienting a specimen for visual examination and, especially, for photomicrography. At least one currently manufactured square stage also rotates.

Stages on polarizing microscopes not only rotate, but they are also graduated in degrees—an essential feature for work with crystals, i.e., when measuring interfacial angles and extinction angles. The type of rotating stage depends on its quality and on whether or not the stage is graduated. An inexpensive rotating stage may ride only on a

layer of grease. More expensive stages rotate on a single row of ball bearings; other stages ride on two rows of ball bearings and rotate with great precision. Graduated stages are provided with a vernier that can be read to a tenth of a degree. Some stages are centerable; others are not, depending upon whether the objectives or objective nosepiece can be centered.

Glide stages are glass or metal plates that slide on a layer of grease. Many square stages and some rotating, circular stages have built-in mechanical stages. Depending on the design, this may be more of a bother than an attachable mechanical stage. The least desirable stage is a fixed square; the most desirable is a circular graduated stage that rotates on ball bearings.

Other features of some graduated, rotating stages include a stage rotation arrest or lock, a slow-motion control to the rotation, and a device that can be set to allow click stops at 45° intervals of rotation. Some stages have openings where the condenser will be oiled to the bottom of the slide. Others have a removable metal plate in the center.

Mechanical Stage

The mechanical stage is one of the most useful accessories for a compound microscope. The stage is a device both for holding a specimen slide firmly in position and for moving it smoothly, either back and forth or right and left. This feature is of particular importance since it enables the user to scan a slide easily in order to locate a suitable field for photography. Then, if the mechanical stage is graduated, it becomes possible to note the graduation figures for the position of a specific field for future reference, in case the same field must be relocated and, possibly, rephotographed.

A mechanical stage is not always included with a microscope, so it is necessary for the buyer to specify that one be included. If a microscope in use does not include this type of stage, it is often possible to purchase an attachable stage separately.

In the absence of a mechanical stage, however, one should not overlook the simple but highly useful stage clips that are normally supplied with a microscope. Stage clips should always be used to hold a specimen slide in place during photomicrography when a mechanical stage is not available.

Field Finder

In addition to a graduated mechanical stage, another device for accurately relocating fields on a specimen slide is called a *field finder*.

This is a microscope slide containing a grid with rectangular coordinates, which are numerically identified.

When a particular field has been located on a specimen slide, the slide is carefully removed from the mechanical stage and the field-finder slide is put in its place. The coordinates for the visible field are then recorded for future reference. Whenever the original field must be relocated, the field-finder slide is placed on the stage and the stage is adjusted to the recorded coordinates. The specimen slide is then substituted for the field finder, and the original field is visible.

If a graduated mechanical stage is used to locate and relocate fields, the technique is satisfactory as long as the same microscope or one with an identical stage is used. Graduated stages, however, vary from one manufacturer to another, and often between microscopes of the same manufacture. The field-finder slide provides the advantage of being interchangeable between microscopes. Field-finder slides can be purchased either from biological supply firms or microscope manufacturers. They go by special names, such as Maltwood finder, Lovins finder, England finder.

Several other methods exist for refinding the same field of view which involve marking the specimen slide itself. One of these methods utilizes a so-called *object marker*. This is a device shaped like an objective which screws into the nosepiece like an objective, but which has a spring-loaded diamond to scribe a circle on the glass. A similar object marker uses a rubber stamp "o" that leaves a circle of ink on the cover glass.

An inexpensive way of marking a field for photography is simply to cut out triangular arrows from paper and glue them on the cover glass with the point immediately next to and above the specimen area of interest.

Accessories

Reticles or graticules are engraved glass discs or glass-mounted microphotographs that lie on the eyepiece diaphragm, where they are brought into focus with the focusing eye lens. This allows one to superimpose a scale, net, protractor, or cross hairs on to the object image so that measurements can be made.

Stage micrometers are ruled glass or metal plates, or reduced photographic scales on microscope slides of known, accurate length used to calibrate eyepiece micrometer scales.

Light filters are essential for use in both visual microscopy and photomicrography. They are used to (1) adjust the color balance of the light to the best visual or photographic condition, (2) absorb heat, (3) reduce light.

Light filters are essential for use in both visual microscopy and photomicrography. They are used to

- Adjust the color balance of the light to the best visual or photographic condition,
- Absorb heat,
- Reduce light intensity,
- Increase contrast,
- Provide monochromatic light for special purposes, such as refractive index determination, transmission interference measurements, and phase contrast work.

These filters may be gelatin, or they may be solid glass. Their use is discussed in detail on pages 49 and 59.

A phase telescope provides an enlarged view of the objective back focal plane for the purpose of aligning the condenser annulus and the objective phase plate ring. It is not essential since the objective back focal plane may be seen by simply looking down the microscope tube after removing the eyepiece; but it is convenient because of the enlarged image obtained with it. The built-in Bertrand lens in polarizing microscopes also provides an enlarged view of the objective back focal plane and can be used to align phase-contrast systems.

Magnification

In photomicrography, image size (or magnification) is controlled by the magnifying powers of the objective and eyepiece and by the bellows extension (or eyepiece-to-film distance). When you look into a microscope, the visual magnification is equal to the product of the magnifying powers of the objective and eyepiece. When the microimage is projected from the eyepiece, this magnification is reproduced at a distance of 250 millimetres (10 inches) above the exit pupil of the eyepiece. If the distance is greater than 250 mm, magnification will be increased proportionately; if the distance is less than 250 mm, magnification will be decreased proportionately. For example, if the objective is designated as 20X and the eyepiece as 10X, visual magnification in the microscope will be 200X. Then, if the image is projected to a ground-glass screen (or film plane) 500 mm (20 inches) above the eyepiece, magnification will be 400X. But, if the projection distance is only 125 mm (5 inches), magnification will be only 100X.

The magnification as engraved on objectives is the *nominal* magnification. The actual magnification is almost always slightly different. For example, a 10X objective may actually have a value between 9.2X and 11.4X, and a 20X objective may actually have

a value between 18.4X and 21.8X.* When the exact total magnification is required, or when the magnifying power of either the objective or the eyepiece is unknown, or when the exact eyepiece-to-film distance is unknown, it becomes necessary to measure magnification. This can be done by means of a calibrated scale called a *stage micrometer*. Such a scale consists of finely ruled lines on a microscope slide in decimal parts of either inches or millimetres. In use, the stage micrometer is placed on the microscope stage and its magnified image is projected to the ground glass or film plane of the camera. By measuring the separation of lines on the ground glass and comparing it with the original separation on the slide, image size can be measured directly. When no ground glass is used, the image of the lines can be recorded on film and measured there after the film is processed. Total magnification can be carried through to an enlarged print or projected slide image.

It is sometimes desirable to record a micrometer scale image simultaneously with a specimen image. This can be done in either of two ways. The stage micrometer image and the specimen image can be recorded together by double exposure: first one image is exposed, then the other, on the same film. In this case, it is advantageous to have a micrometer scale with white lines on a dark background. Another technique is to use a focusing eyepiece with a micrometer scale included. In this case, the micrometer scale in the eyepiece is calibrated with a stage micrometer beforehand so that the line separations will be known. Then, both eyepiece scale and specimen image are recorded in one exposure. The calibration of the eyepiece micrometer must be repeated with every change of film distance or objective.

If magnification in the microscope is carried beyond the point of resolution, no more detail is resolved. This condition is called *empty magnification*. The NA of an objective can be used to estimate maximum useful magnification. Multiply the NA value by 1000 to obtain a figure that, although not exact, is close enough for practical purposes. Judgment of useful magnification involves visual acuity, which varies among observers and can change the value by a factor of 2. An NA of 0.65, then, will allow useful magnification up to 650X. If more magnification and better resolving power are wanted, use a lens with a higher NA (see Fig. 1-9).

The 1000X NA rule to determine maximum useful magnification was the basis for the following table:

---

*Actually observed values, not manufacturing tolerances.

Common Objectives		Initial Magnification	Numerical Aperture	Working Distance	Useful Magnification
Achromat	(40 mm)	3X	0.12	35 mm	120X
Achromat	(16 mm)	10X	0.25	16 mm	250X
Apochromat	(8 mm)	20X	0.65	0.7 mm	650X
Apochromat	(4 mm)	40X	0.95	0.12 mm	1000X
Apochromat	(2 mm)	90X	1.30	0.11 mm	1300X

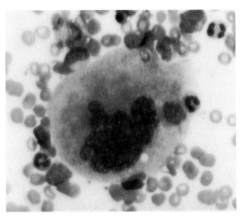

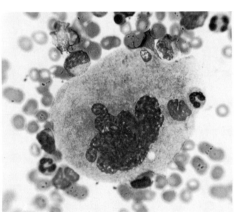

Fig. 1–9
EMPTY MAGNIFICATION— *The image at the top is insufficiently sharp because it was made with a low-power, low-NA objective and subsequently enlarged for reproduction. The image below was made to sufficient size for reproduction with an objective of high NA and high magnification.*

The last column shows the magnification beyond which no further detail is revealed. It is interesting because it suggests that the maximum useful magnification of the light microscope is 1300X, or perhaps 2000X at the very limit. Many research microscopes using 100X objectives, 25X eyepieces, and a tube factor setting of 2 easily achieve magnification of 5000X, but the fact is that no more detail is revealed at this magnification than at 1500X. One simply gets a larger, fuzzy image! See Figure 1-9 again. No matter how many more times Figure 1-9a is magnified, it will not show any more detail than it already has.*

In the last century, before the concept of numerical aperture had been discovered and formalized, magnification was thought to be the sole criterion for revealing detail. Many a time the story is told of 500X objectives being made and microscopes placed on top of other microscopes to enlarge their images still more. Incredibly high magnifications were achieved without revealing any more detail than was revealed with a magnification equal to 1000X NA of the initial objective. This mistake is still made today, and questions as to the highest power of a microscope demonstrate ignorance of the roles of numerical aperture and wavelength in revealing specimen detail. Review the discussion of diffraction theory on page 9.

Depth of Field and Depth of Focus

In ordinary photography, depth of field is considered as the distance from the nearest object plane to the farthest object plane in acceptable focus. When objects are a considerable distance from the lens, depth of field is large, but with objects closer to the lens, the depth of field decreases rapidly. At closest focus with the normal camera lens, the object-to-lens distance is still several times the lens-to-image distance.

In photomicrography, as in microscopy, the object-to-lens distance is considerably less than the lens-to-image distance. Depth of field is exceedingly short and is expressed in micrometres. The equation for photomicrographic depth of field, d, is

$$d = \frac{\lambda \sqrt{n^2 - (NA)^2}}{(NA)^2}$$

where λ is the wavelength of light being used, n is the refractive index of the medium between objective front lens and specimen slide, and NA is the numerical aperture. λ and d must be expressed in the same units. Thus, if d is desired in micrometres, λ must be expressed in micrometers (white light may be taken at $0.55\mu m$).

Now this is a very interesting equation. It shows how much depth within the specimen plane will be in sharp focus at any one focus setting. Generally speaking, you would like depth of field to be large, as it is already going to be quite shallow just from the nature of a system in which the lens is brought very close to the subject. If you want d to be large, notice that λ should be large, i.e., toward the red end of the spectrum (but remember that this is the exact *opposite* of what is required for maximum resolution). Note, too, that n should be large, i.e., immersion systems are suggested. This is useful because high n favors high NA, which in turn is required for good resolving power. But observe the effect of the value of NA on the formula. Taking the square root of the difference in the squares reduces the effect of $(NA)^2$ in the numerator, leaving the $(NA)^2$ in the denominator to have a profound effect on d. Specifically, as NA increases, d rapidly decreases. In other words, large depth of field is favored by *low* NA. But recall that this is the exact opposite of what is required for maximum resolution. Since it has already been indicated that the setting of the aperture diaphragm in the condenser controls the NA on which both the resolving power and depth of field depend, the correct use of this diaphragm is of the utmost importance. In practice, the photomicrographer adjusts the aperture diaphragm to achieve a compromise

*The 1000X NA limit is based on Abbe's estimate of visual acuity. Several workers regard this as an overestimation and consider 1000X NA as the *minimum* for most people (assuming a viewing distance of 250 mm). They regard 2000X NA or even 4000X NA as comfortable and therefore not empty.

between resolving power and depth of field. The proper setting of this diaphragm is discussed in the section on illumination.

The following tabulation shows the approximate depth-of-field values for dry objectives of various numerical apertures. Depth of field also decreases as wavelength decreases, so computation for only one wavelength, green, is shown here.

Variation in Depth of Field with Change in NA (green light)

NA	0.25	0.30	0.50	0.65	0.85	0.95
Depth (in μm)	8.52	5.83	1.91	0.99	0.40	0.19

It is clear from the table that depth of field is extremely small for objectives of high NA. Focus of the image with objectives of high NA becomes very critical, and just a touch of the fine-focus adjustment on the microscope may cause the image to go out of focus. This very limited field depth also presents a problem when the specimen is too thick for the objective in use. The out-of-focus areas within the specimen will scatter and diffuse the light passing through, affecting the recorded image quality in photomicrography. Ideally, the specimen should be no thicker than the usable depth of field for a given objective, but this is seldom achieved. For instance, the average histological section may be around 8 μm thick, yet the greatest depth of field with, say, the 40X/0.65 objective is about 1 μm. Thus, the section in this example has 7 μm of out-of-focus image. Notice in the tabulation that an objective of NA 0.25 would have a depth of field of about 8 μm. So, the aperture diaphragm could be closed down until the effective NA of the system was reduced to 0.25, in which case the entire thickness of the section could be recorded. Note that reducing the NA from 0.85 to 0.25 severely cuts down on the resolving power.

The problem is not so acute when the specimen is examined visually in the microscope, since the eye can rapidly accommodate for the actual depth. Then, too, the microscopist constantly changes focus from top to bottom of the specimen in order to see the entire structure. Visual depth of field is apparently greater than photomicrographic depth of field due to eye accommodation. Visual field depth (in millimetres) is:

$$\frac{250}{M^2}$$

where M is the total microscope magnifica-

tion for those able to accommodate for an object distance of 250 mm.

In photomicrography, however, the image is recorded on the film in one plane. Any pronounced difference between depth of field and specimen thickness may affect the quality of the recorded image.

Be careful not to confuse depth of field with depth of focus. Depth of field, which is what we have been discussing, is in the *plane of the specimen*. Depth of focus is in the *plane of the film*; it is the depth of the image in the film plane that is in acceptable focus. Depth of focus (d′) is related to depth of field (d), in the following way:

$$d' = d \times M^2$$

where M is the total microscope magnification.

This equation holds a surprise. If you substitute the appropriate values for, say, a 4X/0.16 objective and a 100X/1.3 objective, and use both with a 10X eyepiece, you will find that the 100X oil-immersion objective gives a *greater* depth of focus than the 4X objective, even though its depth of field is much less. This fact is important to remember when checking a photomicrographic unit for focus correspondence between the visual image and the film plane. See page 43.

While small errors in focus setting are not apparent in photomicrographs made with high power objectives because of relatively large depth of focus, the same errors may be readily apparent in low power photomicrographs. The table gives an indication of the relative magnitude of the image depth at various magnifications.

Eliminating Spherical Aberration

Selection of Cover Glass

Most objectives intended for use with transmitted light must be used with *covered* specimens; that is, a *cover glass* must be mounted over the specimen. The thickness of the cover glass is specified as either 0.17 mm or 0.18 mm, depending on the manufacturer. The specification for cover-glass thickness results from the fact that objectives are corrected for spherical aberration only when used with a cover of the right thickness. Deviation from this thickness can result in appreciable overcorrection or undercorrection for spherical aberration, particularly with high-aperture, dry objectives. Either condition will result in poor photomicrographs.

The quality and thickness of the cover glass are controllable factors of great importance for the best image quality. For best results in photomicrography, cover glasses should be scrupulously clean and within a thickness range of 0.16 mm to 0.19 mm (No. 1½). One way to prevent the problem of incorrect cover glass is to prepare the specimen properly. This is, however, not as easy as it sounds. The objective sees not just the cover glass but the mounting medium above the specimen, and perhaps even a part of the specimen itself. That is, the objective "sees" the total optical path length of the same refractive index from specimen to top of cover glass. Mounting the specimen directly on a measured cover glass, and then to a slide, reduces the amount of

Image Depth in the Intermediate Image Plane as a Function of Objective Magnification

Objective Magnification	Image Depth (mm)		
	5 μm specimen	8 μm specimen	10 μm specimen
2.5X	0.031	0.050	0.063
4X	0.080	0.128	0.160
6.3X	0.198	0.318	0.397
10X	0.50	0.80	1.00
16X	1.28	2.05	2.56
25X	3.13	5.00	6.25
40X	8.00	12.8	16.0
63X	19.8	31.8	39.7
100X	50.0	80.0	100.0

Selecting Cover Glass Thickness

NA of Objective	Required Thickness (mm) Continental Objectives	U.S. Objectives
0.07–0.30	Use with or without cover glass.	
0.50–0.65 (and all oil immersion)	0.17 ± 0.05	0.18 ± 0.05
0.70–0.95	0.17 ± 0.01	0.18 ± 0.01

mounting medium in the optical path. It is possible to mount specimens in this way, but is not commonly done. It is safer to assume that there will be some abberation introduced even with measured cover glasses and to concentrate on ridding the system of it.

What changes in cover-glass thickness are significant? And how can you correct for aberrations introduced by incorrect cover-glass thickness? Objectives with NA 0.30 or less (typically objectives of 10X or less) can be used with or without a cover glass. Objectives with NA 0.50 to 0.65, and all immersion objectives, should be used with cover glasses no more than ± 0.05 mm from the thickness recommended by the manufacturer. *High-dry* objectives, i.e., those with numerical apertures of 0.70 through 0.95, are the most sensitive to incorrect cover-glass thickness. Objectives with NA 0.70 to 0.95 should be used with cover glasses that differ by no more than ± 0.01 mm from the thickness recommended by the objective manufacturer. Objectives in this sensitive range will usually have the recommended cover-glass thickness engraved right on the objective's barrel along with the magnification and numerical aperture, e.g., 40X/NA 0.95/0.17. Generally speaking, U.S. and British objectives are corrected for cover glasses 0.18 mm thick, and continental and Asiatic objectives are corrected for use with cover glasses 0.17 mm. There are exceptions, so that one should always read the legend on the objective in use. A machinist's micrometer of either the drum type or the dial type, in English or metric units, serves to measure individual cover glasses for critical use.

With a layer of mounting medium between the specimen and bottom of the cover glass, some spherical aberration is bound to be introduced. Spherical aberration in images in the 400X to 600X range are commonly seen in the published literature. Photomicrography at higher and lower magnification is comparatively much easier. Photomicrography with high-dry objectives is most difficult. But it need not be. It only remains to cancel out the introduced spherical aberration by deliberately introducing more spherical aberration of the opposite sign and magnitude.

The three major ways of eliminating spher-ical aberration due to too thick or too thin cover glasses are (1) to alter the mechanical tube length, (2) to alter the optics, or (3) to make use of immersion objectives in the 40X to 63X range.

Mechanical Tube Length Method

Mechanically adjusting the tube length varies the distance between the objective and eyepiece. Whether or not this method can be used depends on the kind of body tube one has on the microscope. See the discussion of **Tube Length,** page 10.

If the microscope is equipped with a drawtube, it is normally set to either 160 mm or 170 mm, depending on the manufacturer. If now a photomicrograph needs to be made of a specimen using a high-NA, dry objective, proceed as follows: Focus carefully on some fine detail within the specimen—the granules in leucocytes in a blood film, for example. Now, arbitrarily change the drawtube length a few millimetres either way. First you will have to change the focus slightly because the focus and magnification will have been slightly changed. Then evaluate the image; i.e., ask yourself whether the change in mechanical tube length has improved the quality and clarity of the image. If there was an improvement, continue to shorten or lengthen the tube in the same direction that effected the improvement and repeat the evaluation. If the initial change resulted in image deterioration, proceed in the opposite direction. Continue to change the tube length by trial and error in this manner until the best quality image is found. This will be the point where the system is free of spherical aberration, and a high-quality photomicrograph will result. *Unfortunately, most modern microscopes do not have a drawtube so that this convenient and inexpensive method cannot be used.*

Even with a drawtube, this method doesn't always work because if the cover glass is of incorrect thickness, chances are it will be too thick rather than too thin. And too thick covers are compensated for by closing, i.e., shortening the drawtube. The closed position on many drawtubes may be restricted and prevent the tube from being closed enough to effect enough correction. Usually there is plenty of draw for too thin covers. In fact, too thin covers can sometimes be corrected even on microscopes with rigid body tubes by drawing the entire eyepiece out a bit, thereby effectively increasing the tube length. In the absence of a drawtube, one of the following methods must be used.

Optical Methods

One of the best methods of eliminating spherical aberration due to incorrect cover-glass thickness is through the use of objectives equipped with *correction collars.*

Because the majority of specimens are *not* mounted directly to precisely measured cover glasses, microscope manufacturers provide high-dry objectives (NA 0.85 to 0.95) with correction collars. A spring is placed between the front lens and the rear lens systems instead of a fixed spacer. A knurled ring around the outside of the objective can be turned to vary the distance between the lenses. The ring usually bears a fiducial mark, and a scale is provided indicating actual cover-glass thickness (0.11 to 0.23 mm) or arbitrarily numbered divisions. In use, you should set the mark at mid-range and focus on some fine detail in the specimen. With one hand on the objective correction collar and the other on the fine-focus adjustment, turn the collar slightly in either direction, touch up the fine focus, and compare the image quality with the first setting. If the image improves, continue to change the collar in the same direction. If the image deteriorates, turn the collar in the opposite direction. Continue this process of changing the correction collar setting, adjusting fine focus, and image evaluation until no further improvement in image quality results. Only then make the photomicrograph.*

Immersion Objective Method

Of all the methods to eliminate spherical aberration due to too thick or too thin cover glasses, the immersion objective method is the best. It will be recalled that the spherical aberration is due to either too long or too short a light path through specimen, mounting medium, and glass before refrac-

---

*The Jackson tube-length corrector, once made by a British firm, served the same purpose as the built-in correction collar. The Jackson tube-length corrector consisted of a lens system that fit between the objective and the body tube. A knurled ring changed the internal lens position, thereby changing the tube length optically.

tion takes place at the glass-air interface. By replacing the airspace with an immersion oil that has about the same refractive index as the glass and mounting medium, the glass-air interface refraction is eliminated, and along with it, the problem itself. Use of immersion objectives is described in the next section.

Cover-glass thickness is not as critical with oil-immersion objectives when the refractive index of the oil is nearly equal to that of the cover glass. This refractive index may vary, however. With immersion objectives of highest aperture (NA 1.2 and 1.4), it is still important to use cover glasses within the ideal thickness range. While cover glasses of other-than-specified thickness can be utilized in visual microscopical examinations, it is good practice to use a cover glass of the correct thickness in case a photomicrograph is required.

Objectives intended for use with reflected light require special objectives for uncovered objects. No cover glass is used over the specimen. Objectives of this type may be marked with a dash (−) where cover glass thickness is ordinarily indicated, or they may be marked *o.d.* (German for *ohne decke*—without cover). This principle applies mainly to metallographic microscopes and metal specimens, and their use is described in the section on metallography, page 80.

Still another series of objectives are corrected for *infinite tube length,* rather than some fixed distance. Objectives of this type will be marked with an infinity symbol (∞).

Use of Immersion Objectives

To achieve high magnification *and* high resolving power in a photomicrographic system, it is necessary to use immersion objectives. Recall from the theoretical consideration discussed in the **Numerical Aperture** section that a dry system is limited to a practical numerical aperture of 0.95, and that this places a limit on the ultimate resolving power. Resolution of fine specimen details requires that the minimum numerical aperture in the condenser-specimen-objective system be 1.00 or greater. This is achieved by placing a liquid between the specimen cover glass and the front lens of the objective and, for NAs greater than 1.00, between the top lens of the condenser and the bottom of the slide.

Immersion objectives in the 40X to 63X range are not used as often as they should be. There are probably three reasons for this:

(1) ignorance of their benefit, (2) expense, and (3) the additional technique involved. Immersion objectives in the 40X to 63X range have several benefits over high-dry objectives. They are usually of higher correction (fluorites or apochromats), and they almost always have higher numerical apertures (1.0 to 1.4) compared to high-dry objectives (0.65 to 0.95).

Immersion objectives can be used with larger aperture diaphragm openings, yielding greater actual utilization of numerical aperture. Immersion objectives are, however, more expensive than high-dry objectives without correction collars. Many microscopists regard immersion systems as messy and a bother. It is true that the use of immersion systems requires more time and care, including cleanup afterward. For the highest quality photomicrography, the additional time and technique is well invested.

The upper theoretical limit of the numerical aperture for immersion systems is the refractive index of the liquid with which the objective is used. Immersion systems have been computed for water (n = 1.33), glycerin (n = 1.47), cedarwood oil or synthetic immersion oil (n = 1.515), and monobromonaphthalene (n = 1.66). The most common of these is oil immersion. Objectives intended for oil immersion will be marked variously— *oil, oel, HI* (for homogeneous immersion)—or will simply be marked with an NA of 1.0 or greater. Water immersion will be similarly marked *water, water immersion, wasser,* or *WI*.

Immersion oil can be obtained from biological supply firms, science supply houses, and microscope manufacturers or their dealers. For critical use, the dispersion (as well as the refractive index) of the oil should be that recommended by the manufacturer. It is best to use oil supplied by the manufacturer of the objective, unless the dispersion is known and can be matched. In the past, many immersion oils were made with polychlorinated biphenyls (PCBs). Although these are no longer made, extreme care should be taken to avoid skin contact or ingestion if they are encountered. Oil immersion objectives should not be used with glycerin or other media for which they were not designed.

A bottle of immersion oil should never be shaken, since air bubbles may be introduced. A single, microscopic bubble under the objective lens will cause flare, which will lower contrast and affect image quality. Also, the bottle should never be left open to the air, since dust may settle on the oil. Dust or other

---

*See **Diffraction Theory and Resolution** page 9.

debris in the oil will impair image quality.

When the NA of the objective exceeds 1.00, oil should be placed between the condenser top lens and bottom of the slide. If oil is omitted here, the system NA will not exceed 1.0 regardless of what is engraved on the objective. Common 100X oil immersion objectives have an NA of about 1.25 (up to 1.40). The objectives are almost always used oiled to the top of the cover glass, but the slide is seldom oiled to the condenser. It must be realized that under these circumstances the NA is not 1.25 as engraved, but 1.0 or less. Only if the condenser is oiled to the slide will the full NA and resolving power be achieved. The only possible excuse for omitting oil from the condenser top lens is that most 100X objectives in use are achromatic and almost always require some closing of the aperture diaphragm, which will reduce the NA anyway. If this is the reason, it is excusable—as long as one realizes the resolving power limitation.

To achieve full resolving power and NA from objectives with NA greater than 1.0, oil the condenser to the bottom of the slide before attempting to focus the objective. (If the objective has NA less than 1.0, start with Step 6 below.) Follow this sequence:

1. Make sure that the condenser is made for immersion use. If it is not engraved with an NA of 1.00 or greater, *do not* immerse it.
2. Remove the specimen slide from the stage and place a drop of oil on the top lens of the condenser.
3. Rack the condenser down slightly.
4. Place another drop of oil on the bottom of the slide under the specimen.
5. Rack the condenser up until the oil on the condenser top lens contacts the oil on the bottom of the slide.

(Start here if the condenser is not being oiled to the bottom of the slide.)

6. Find a suitable field on the slide with a low-power (dry, 10-25X) objective. Locate the area of interest in the very center of the field. Focus the substage condenser to image sharply the field diaphragm in the plane of the specimen (as described in the section on setting up correct illumination according to Köhler).
7. Raise the body tube or lower the stage using the coarse focus control and move the oil-immersion objective into position. (Some oil immersion objectives are made so that the lower half can be retracted up into the upper half, making it unnecessary to move the coarse focus.)
8. Place a small drop of oil on the front of

the objective and another small drop on the cover glass over the area of interest. This technique will prevent formation of air bubbles. The drop of oil on the front of the objective is frequently omitted, but this is a mistake, particularly with many modern plano objectives that have concave front lenses. Omitting the oil droplet on the objective front lens is asking for entrapped air and the resulting flare.

9. Using the coarse adjustment, bring the objective lens and specimen slowly together until the oil on the lens contacts the oil on the cover glass. A flash of light will occur at this moment, visible when the eye is close to the level of the lens. DO NOT PERFORM THIS STEP WHILE LOOKING INTO THE MICROSCOPE. Always observe the oil contact from stage level. Now, while looking into the microscope CARE-FULLY continue to bring the objective and specimen together with the coarse adjustment while watching constantly for specimen focus.

10. Now use the fine adjustment on the micro-scope for final focus of the specimen. (It is poor technique to resort to the fine adjust-ment immediately after making oil contact. The distance is likely to be so great that the fine adjustment will run to the end of its travel without ever getting the specimen in focus.)

The careful use of immersion objectives is one of the techniques in microscopy that needs to be practiced to achieve perfection. Carelessness at this point may result in damage to specimen or objective. Immersion objectives have very short working distances (as small as 0.09 mm).

Many microscope users end up with oil all over the stage from moving the slide around. The oil on the bottom of the slide is dragged onto the stage when the specimen is scanned. All microscopes intended for advanced tech-niques have a stage insert that is removable. The opening in the stage is frequently in a removable metal disc. Removal of this stage insert allows scanning of large specimens without getting oil on the stage. In polarizing microscopes, the opening is especially large, as it is made so for use with the universal stage. Some microscope stages are made with large, wide-oval stage openings for use with oiled condensers.

Immersion oils may come in various viscos-ities. For ordinary immersion use with the microscope in its normal, vertical position, the medium viscosity is perfectly adequate.

More viscous oil is used with long-working distance condensers, with horizontal metallur-gical microscopes, or with projection micro-scopes where heat may be a problem. Light oil is used for objectives and condensers with high NA and very short working distance, or in colder environments.

When you are through using the oil-immersion lens, remove the oil. So-called nondrying immersion oils do solidify. If the oil is not cleaned off the objective lens, it will dry and effectively change the curvature of the front lens. When the lens is used later, the image will appear to suffer from severe astig-matism. Gently wipe the front surface with a clean, dry lens tissue then follow by wiping with a tissue dampened with a suitable solvent.* Remove oil from the slide and the condenser, if necessary, in the same manner.

---

*Use the solvent recommended by the objective manu-facturer. When no recommendation is available, use 1,1,1-trichloroethane (Eastman Organic Chemical No. T3613) in preference to benzene, xylene, or related compounds.

Chapter Two
MICROSLIDES

How Microslides Are Prepared

Basically, any transparent specimen to be examined or photographed through a microscope is mounted in a suitable medium on a microscope slide and covered with a cover glass. The specimen is often colored by means of various biological stains to produce contrast and visualization of structural details. Tissue sections and smears are common types of specimens prepared for the microscope although many other subjects can also be suitably prepared.

The actual techniques used in slide preparation can be quite involved and complex, and will not be explained in detail in this book. However, the fundamentals of slide preparation will be reviewed briefly. More detailed explanations can be found in books on microscopy and microtechniques.

A wide selection of prepared microslides is available by mail order from several scientific and biological supply firms. These slides include tissue sections, smears, and many other types of specimens. Prepared slides are used extensively in classroom instruction in biology, botany, zoology, paleontology, geology, mineralogy, petrography, microbiology, medicine, criminalistics, and materials science. See Appendix I.

Slides and Cover Glasses

Normally, a specimen is mounted on a glass slide that is usually 3 inches long and 1 inch wide (76 mm x 25 mm). All brightfield microscopes are equipped to handle slides of these dimensions. Larger slides are also used, although less often than the 1 x 3-inch size. The larger sizes (1½ x 3-inch or 38 x 76 mm and 2 x 3-inch or 51 x 76 mm) will accommodate either large sections or a long series of sections. Standard petrographic and metallographic slides are shorter than biomedical slides (1 x 2-inch or 25 x 51 mm).

The surface of the slide should be flat. Also, the glass should be of high quality and chemically stable (corrosion-resistant) since it may be subjected to many reagents. All commercially available microscope slides are normally produced with these characteristics.

The thickness of the microscope slide is of definite importance when high-power photomicrography is considered because the high numerical aperture condensers necessary for high NA objectives have shorter free-working distance. Condensers are designed for use with slides of specific thickness, within a narrow range of tolerance. Unfortunately, the specifications for slide thickness vary with different microscope manufacturers. One condenser, for example, may require a slide thickness of 1.2 mm, whereas another may specify 1.6 mm. The specification for slide thickness is related to the numerical aperture and working distance of the condenser. Slide thickness is less critical for visual microscopy or for photomicrography at lower power when the top of the condenser is removed. In this case, the condenser has a much longer working distance and the actual slide thickness is unimportant.

If a specimen is to be photographed at high magnification, however, it may be necessary to make a special preparation on a slide of the correct thickness, as specified for the condenser in use. In general, however, a slide thickness of 1.0 mm is acceptable for most subjects to be examined or photographed at both low and high magnifications.

The cover glass is usually a circle, square, or rectangle of very thin optical glass. Various sizes are available for the different shapes, but if a 25 x 76 mm slide is used, the cover glass should be of suitable size to fit the slide. The most common size for a 1 x 3-inch slide is 22 mm, either round or square. Rectangular cover glasses should be less than 25 mm wide. Smaller sizes are used occasionally for small specimens. Plastic cover glasses should not be used for photomicrography.

Ideally, the appropriate thickness of the cover glasses (either 0.17 mm or 0.18 mm) should be as specified by microscope manufacturers. Cover glasses are available in different thickness ranges as No. 0, No. 1, No. 1½, No. 2, and No. 3. Each number indicates a narrow range of thicknesses, with No. 0 being the thinnest and No. 3 the thickest. The complete range is from less than 0.1 mm to

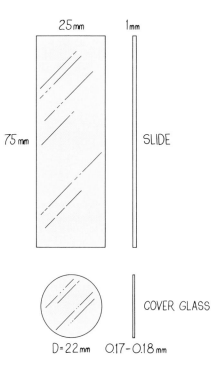

Fig. 2-1
DIMENSIONS OF SLIDES AND COVER GLASS

slightly over 0.3 mm. A box of No. 1½ cover glasses includes the ideal range of 0.16 mm to 0.19. It is therefore best to prepare a microslide with the No. 1½ cover glass. The only exception would be for preparation of some whole mounts, where a thick cover glass (No. 2 or No. 3) would be of advantage. (For a description of whole mounts, see page 22.) When the right thickness of cover glass is

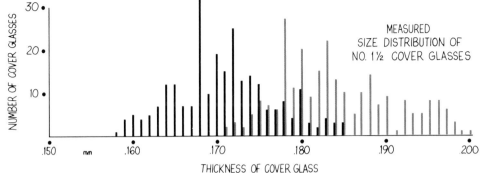

Fig. 2-2
COVER GLASS THICK-NESS—*In an actual analysis of the thickness distribution of No. 1½ cover glasses in samples from two different manufacturers, the result was as shown. The mean thickness for batch A was 0.170, ideal for continental objectives. The mean for sample B was 0.180, ideal for American or British objectives.*

used, the slide can be routinely examined visually and will be correct for most photomicrography, particularly with a high, dry objective. (See the previous discussion on **Selection of Cover Glass**, page 16.)

Tissue Sections

Many of the specimens prepared for the microscope are very thin transparent slices, or *sections*, of biological material, usually either animal or plant tissue. Sections are cut on an instrument known as a *microtome*. A block of tissue is held firmly in a clamp and passed over the edge of a very sharp, stationary knife blade to produce a thin slice of the material. The thickness of the slice is controlled by a setting on the microtome. Serial sections, one after the other, of the same thickness can be cut in this way.

For larger pieces of tissue, a *sliding microtome*, where the block remains stationary and the knife moves back and forth to cut thin slices, is used. A very large and heavy version of sliding microtome is the *sledge microtome* with a massive bed which holds the specimen stationary while the knife is pushed or pulled over the specimen in a planing action. This microtome is essential for cutting harder specimens, such as wood.

Most tissue specimens, however, are not hard enough to be placed in the microtome and be cut without preliminary preparation. The tissue must first be fixed and hardened with a chemical reagent. Then it is dehydrated, cleared, and placed in a bath of embedding matrix such as molten paraffin wax where it is left long enough for the wax to impregnate the tissue. The impregnated tissue is then embedded in a larger block of paraffin. This is accomplished by placing the impregnated tissue in a box-like mold containing melted paraffin. The mold is cooled rapidly by immersing it in cold water to harden the paraffin and to prevent crystallization. The result is a block of wax containing the tissue specimen. This block is trimmed to a convenient size and attached to a holder that is inserted in the microtome clamp.

Sections are cut on a microtome either individually or as ribbons. A ribbon is a long chain of sections that cling to each other at the edges. Either one section or a short ribbon is placed on a microscope slide that has been treated with an adhesive material. The embedding medium is then removed by bathing the slide in a suitable solvent. The sections fastened to the slide are usually colorless and practically transparent, unless stained (see below) before embedding. Little or no detail would be visible if they were examined in an ordinary microscope. In order to render the details visible, the sections can be colored with a *stain*; many different biological stains are used for the purpose. *Double staining*—that is, the use of two selective stains—is common in order to differentiate between specific parts of a specimen. For example, with animal tissue the nucleus of a cell might become stained with one color and a different stain might color the cytoplasm, the area of the cell around the nucleus. The most common stains used for animal tissue sections are hematoxylin and eosin. These stains appear as blue and light red, respectively. (See Kodak Publication No. JJ-281, *EASTMAN Biological Stains and Related Products*.)

Many other colored stains are also used in both plant and animal histology. Some techniques utilize triple, quadruple, and even quintuple staining. Single stains are used rarely.

When tissue sections are to be prepared, it would be best for photomicrography if they were as thin and as well stained as possible. Thick sections present a problem when photomicrographs are to be made at medium-to-high magnification. There are limits in the preparation of thin specimens by the paraffin technique. Sections thinner than 4 or 5 μm are difficult to make. However, sections as thin as 1 or 2 μm are needed for high-magnification work, both visually and photographically.

To make ultrathin tissue sections, it is necessary to prepare them differently and to use an impregnating and embedding substance other than paraffin. Commonly employed in place of paraffin are the methacrylate and epoxy resins. Sections are cut on an ultramicrotome that has either a glass or a diamond cutting edge. This technique is commonly used in preparing sections for electron microscopes.

Another technique for sectioning tissue for the light microscope involves quick freezing. Tissues can be hardened sufficiently by freezing so that the usual impregnating and embedding techniques are eliminated. The material to be sectioned is usually fresh and may be either fixed or left unfixed. The principal advantage is speed since the steps of freezing and sectioning can be carried out in minutes—compared with hours for the paraffin technique. The disadvantages are relatively thick sections, less clarity of detail, and the inability to handle large pieces of tissue because they do not freeze entirely throughout. In pathology, however, the ability to make stained, mounted specimens of human tissue in a very short time is a definite advantage.

No matter how a section is made, it is mounted on a microscope slide, stained, topped with a drop of suitable mounting medium, and covered with a cover glass. The microslide is then ready for viewing or for photography through the microscope.

Smears

Preparing certain types of specimens by *smearing* is relatively simple compared with making tissue sections. A sample of fluid material—such as blood, a bacterial culture, or an exudate—is spread in a thin layer on a clean microscope slide. The smear is then dried and fixed. A mounting medium is applied and the specimen is covered with a cover glass.

The methods of smearing vary slightly for different types of specimens. Blood smears are the most common and are made routinely in hematology laboratories. Two methods of making blood smears are practiced. In the first method, a drop of blood is placed on a clean slide about 1 to 2 cm from the end. A second slide, held at about 45° and in contact with the first slide, is allowed to touch the blood so that capillary action distributes the blood along the trailing edge of the slide. The slide is then pushed forward quickly to produce a thin smear on the bottom slide. The smear is allowed to dry in air and it is then ready for fixing.

The second method for blood smears involves placing a drop on a clean square cover glass. A second clean cover glass is touched to the drop of blood on the first and then dropped diagonally across the first glass. The blood spreads quickly between the two glasses. The two cover glasses are separated immediately, by grasping two projecting corners and pulling with a smooth motion. Each cover glass then contains a smear that can be dried in air, fixed, and mounted on a clean slide. Wright's stain, a combination of methylene blue and eosin dissolved in methyl alcohol, is commonly used for fixing smears in the United States.

Preparation of bacterial smears and smears from exudates is even simpler. A drop of dilute liquid material is picked up with a small wire loop. The drop is then smeared on a clean slide by moving the loop either back and forth or in a circular motion, in contact with the slide. The smear is then dried and stained, and a mounting medium and cover glass are applied.

Whole Mounts

There are many specimens that are mounted directly onto microscope slides in specific media without resort to either sectioning or smearing. These include small insects, protozoa, crustacea, pollen grains, and fibers. The techniques of preparation and mounting are quite diverse. The reader is therefore directed to literature on microtechniques, found in most scientific and biological libraries. Many types of whole mounts can also be purchased as prepared slides from supply firms.

Mounting Media

Microspecimens for a transmitted light microscope are nearly always mounted in some kind of medium, on a slide and under a cover glass. The purpose and characteristics of the medium vary, depending on the type of specimen. Mounting media basically are divided into two classes—those suitable for

permanent mounts and those suitable for temporary mounts.

For stained tissue sections and smears, the mountant is usually permanent; it protects and preserves the specimen for future study or photography. The medium must be reasonably colorless so as to produce a neutral background and not degrade the specimen colors. A mounting medium should have a refractive index fairly close to that of the specimen in order to produce the highest degree of transparency. It should not chemically affect the slide glass or the specimen and should not cause stains to fade. Also, it must adhere to glass.

A permanent mounting medium is usually either a natural or synthetic resin. Canada balsam is probably the most well-known example of a natural resin. It has been used as a mounting medium for over a hundred years, and is still commercially available from many sources. Slides prepared with this medium, however, tend to become yellowish with age if the cover glass is not sealed at the edge to prevent oxidation. Specimen slides may also appear slightly yellow when an excess of the medium is used. Some fading of certain stain colors also occurs after a period of time in Canada balsam that contains excess acid.

Synthetic media are employed quite extensively and are available under a variety of trade names. They are in some ways superior to balsam and are often used as substitutes. The table lists some of the common mounting media.

Occasionally, a permanent medium is selected that has a higher refractive index than either the glass or the specimen. The aim is to make the specimen more visible since it may be colorless (unstained) and practically transparent. The greater the difference in refractive index between the medium and the specimen, the higher will be the visual contrast. This change of contrast with difference of refractive index is called the *visibility index*. To calculate the visibility index, take the numerical difference between the refractive index of the specimen and the refractive index of the mounting medium and multiply by 100. The resulting number can be compared directly to any other visibility index. For example, suppose a siliceous specimen with a refractive index of 1.43 is mounted in Canada balsam (n = 1.53); its visibility index is 1.53 - 1.43 = 0.10; 0.10 x 100 = 10. Now, if the same specimen (n = 1.43) is mounted in, say, Aroclor resin (n = 1.66), its visibility index will be 1.66 - 1.43 = 0.23; 0.23 x 100 = 23. Comparing 10 to 23, you can see that the siliceous specimen will have much more contrast when

mounted in Aroclor resin than when mounted in Canada balsam. One application is in fiber microscopy, especially with animal hairs. The surface texture of wool fibers, for example, is much more visible in a medium of high refractive index than in one where the index is close to that of the specimen. (Some media containing polychlorinated biphenyls—PCBs—may no longer be available.)

When permanent mounts are prepared, a quantity of medium is placed on the specimen and covered with a cover glass. Sufficient pressure must be applied, either by a weight or by a spring-loaded press, to push the cover into place. The cover-glass surface must be parallel to the glass slide, and a minimum amount of mounting medium should remain between the cover and the specimen. The slide must be kept under pressure until the mounting medium has hardened. After hardening, the slide is ready to examine and photograph.

Temporary mounting media are often used with specific subjects for speed and convenience. A slide made with a temporary mount is often discarded after use. Temporary fluid media include water, glycerin, certain oils, corn syrup, and many organic liquids. One precaution that must be observed is that the specimen should not be soluble in the medium or be chemically attacked by it.

Chemical Crystals

The formation and photomicrography of chemical crystals is fascinating, particularly when you use either a polarizing microscope or a conventional brightfield microscope equipped with polarizing elements. Some crystals are *birefringent*, or *anisotropic*, and appear brightly colored when viewed between crossed polarizers. Very striking color photomicrographs can be made of crystals or crystal patterns by using polarizing light.

The preparation of chemical crystals on a microscope slide is relatively easy. The simplest method is by evaporation. Dissolve a small amount of chemical in distilled water (or other solvent) in a test tube or small vial. Place a drop of the solution on a clean microscope slide and allow the solvent to evaporate. Crystals will begin to form in a short time. (Application of low heat, such as holding the slide above the microscope illuminator, will hasten crystallization.) The growth can then be studied under a low-power microscope with crossed polarizers. When all of the solvent has evaporated, crystals can be photographed dry, or a mounting medium can be applied and a cover glass used. The mounting medium can be a permanent type, but its

Common Mounting Media

Mounting Medium	Refractive Index	Use
Canada balsam	1.53	Biological, general
Caedax medium	1.58	General
Diaphane (green) resin	1.54	Especially recommended for hematoxylin-stained specimens
Harleco Synthetic resin	1.52	General
Permount resin	1.54	Biological, general
Aroclor resins*	1.63-1.66	High refractive index work
Carmount 165 resin*	1.65	High refractive index work
Castor oil	1.47	Temporary mounts, general
Cedar oil	1.52	Temporary mounts, general
Immersion oil	1.51	Temporary mounts, general
Corn syrup	1.42	Semipermanent mounts, ultraviolet photomicrography
Water	1.33	Temporary mounts, general
Glycerin	1.47	Temporary mounts, fluorescence work

*The Aroclor resins and Carmount 165 resin may now be difficult or impossible to obtain because they contain PCBs (polychlorinated biphenyls).

Some Chemicals Usable for Making Crystal Patterns

Chemical	Melting Process	Evaporation Process	Comments
Ascorbic acid (vitamin C)		X	Soluble in water; partly soluble in alcohol. Recommend: rubbing alcohol (70% ethyl).
Benzoic acid	X		Melting point: 121.7°C.
Bromo Seltzer		X	Soluble in water.
Citric acid	X	NR*	Melting point: 153°C; evaporation difficult.
KODAK DEKTOL Developer		X	Tricky; sometimes helpful to make sandwich by adding second glass after partial evaporation. **WARNING†**
Dextrose (glucose)	X	NR*	Melting point: 146°C.
KODAK Developer D-76		X	**WARNING†**
Epsomite (epsom salts)		X	Crystals tend to be thick, so experiment with concentration; also try sandwich.
Hydroquinone		X	Variable results; helpful not to dissolve entirely. **WARNING†**
Tartaric acid		X	**WARNING:** Causes eye irritation on contact.
Urea	X	X	Melting point: 132.7°C. Both processes excellent.
Vanillin		X	Soluble in 12 parts H_2O, 2 parts glycerol, and 2 parts 95% alcohol.

*NR—not recommended (less effective)

†**WARNING:** Repeated contact may cause skin irritation and allergic skin reaction. Avoid breathing dust. May be harmful if swallowed. If swallowed, induce vomiting. Call a physician at once. Keep out of the reach of children.

solvent should not attack the crystals. This might happen with some organic chemicals but is unlikely with inorganic materials.

Fusion is another technique that produces colorful crystal patterns under polarized light. Place a very small amount of an organic chemical on a microscope slide and place a cover glass on top. Then apply heat to the bottom of the slide with an alcohol lamp or hot plate. The chemical will melt and spread evenly under the cover glass. When the heat is removed and the melt cools, crystal growth begins when the temperature drops below the melting point. The crystals formed will fuse together, very often producing beautifully colored patterns. Sometimes it happens that the melt is supercooled and no crystallization takes place. In this case, add some of the solid chemical to the edge of the cover glass to seed the melt, causing crystal growth. Some chemicals crystallize quickly, and other crystallize over several minutes or hours.

Only a low-power microscope is necessary for this photography; a 5X or 10X objective will usually suffice. Crystal patterns are often too thick for higher power. One advantage of a fusion preparation is that if you do not like the particular pattern formed, you can remelt the chemical and observe new patterns. The resulting colors between fully crossed polarizers arise from varying thicknesses and different refractive indices of the chemical crystals. One way of changing the colors is to alter the thickness of the preparation either by using more or less of the chemical or, alternatively, by depressing the center or one edge of the cover glass (to form a wedge) while the preparation cools and crystallizes.

If polarizing elements or sheets are being added to a biomedical microscope for chemical crystal photomicrographs, the polarizers should be selected for color. Some polarizing material is decidedly green or amber or gray. The polarizers that are as nearly neutral gray in color as possible will be the best ones for color photomicrography. However, even off-color polarizers can be corrected through the use of color compensating filters, which will be discussed in detail later.

Chapter Three
MICROSCOPE ILLUMINATION

How Microslides Are Illuminated

Correct illumination of the microspecimen is the single most important aspect of critical microscopy and photomicrography. Fully 80 percent of all photomicrographs submitted for contests and exhibitions are rejected because of improper illumination. Another 10 percent have improperly adjusted aperture diaphragms, and since this adjustment is intimately associated with correct illumination, it would be correct to say that 90 percent of all rejections are due to improper illumination and microscope alignment. The same is true of many published photomicrographs including, regrettably, those in many published books of photomicrographs and scientific journals. This point cannot be overemphasized. No matter how well the specimen is prepared, without being properly illuminated the full detail and color of a specimen cannot be realized visually or photographically. Unfortunately, most people have the idea that all one needs is to get light up the tube so that the specimen can be seen and focused upon. This reduces the compound microscope to the level of a hand magnifier, instead of realizing its potential as a critically important scientific instrument. Time spent in learning and practicing the proper method of illumination, as described in this section, will be rewarded by consistent production of successful photomicrographs.

A complete understanding of the principles and practices of obtaining efficient illumination is just as important to the photomicrographer as knowing the capabilities and limitations of the compound microscope. Correct adjustment of the optical system of the microscope is, in fact, dependent upon an efficient system of illumination. An objective, for example, cannot be used effectively unless the substage condenser is properly adjusted and the substage diaphragm is set at the correct aperture. These adjustments are made by following the system known as Köhler illumination. Since the adjustments need to be optimized for each individual specimen, they cannot be set at the factory or once-and-for-all by the user.

The light source itself should provide suffi-cient intensity to allow reasonably short photographic exposure times. The lamp housing should be designed to allow easy access to the light source and should contain those elements necessary for proper adjustment of the illumination furnished to the microscope. It should also be designed to dissipate heat efficiently.

For color-film exposure, the light source should conform to, or allow suitable filtration to meet, the requirements of the film. A source with a continuous visible spectrum is necessary. Most common light sources meet this requirement. One exception is the mercury-vapor lamp, which emits line spectra. This source, however, is of special interest for ultraviolet and fluorescence photomicrography. It can also be used for regular black-and-white work.

Efficient illumination depends upon correct adjustments of the microscope and the illuminator. But it is also very much dependent upon correct alignment of all components of the system—from the light source to the film plane. Many effects of uneven illumination in the image, especially color fringes and hot spots, can be traced to improper alignment. If centering devices are not provided for the condenser and the light source, alignment should already have been established by the manufacturer. The microscope manual or the manufacturer should be consulted if alignment problems persist.

Light Sources

A common light source in general photomicrography is the incandescent tungsten-filament lamp, available in a wide selection of voltages and wattages. Most microscopes with built-in illumination use either a 6- or 12-volt coil-filament lamp. Color temperature varies from less than 2700 K to 3200 K, depending on lamp design, lamp age, and electrical conditions at the time of usage. This low-voltage lamp operates through a transformer that either has several settings at fixed voltages or is continuously variable. Provision is made for overvolting (i.e., running at voltages greater than the nominal maximum value). The highest setting is usually suggested when color film is to be exposed, in order to provide the highest color temperature.* Even when exposing color film balanced for tungsten illumination, it is often necessary to use appropriate light-balancing filters to adjust the illumination to the correct color temperature. The filters required will vary with different lamps. Their proper use is described in the section on light-balancing filters.

When a microscope does not have built-in illumination, an external illuminator must be used. Separate illuminators are available from microscope manufacturers or dealers. These illuminators contain many kinds of lamps including 6- to 12-volt coil-filament lamps, 120-volt coil-filament lamps in wattages from 15 to 100, 120-volt or low-voltage tungsten-halogen lamps, and the 6-volt, 18-ampere ribbon-filament lamp. Here also the illumination must be adjusted with appropriate filters to suit the color film in use.

A problem with tungsten lamps is that the glass envelope becomes darkened with age. The darkening is due to the deposit of tungsten resulting from vaporization of the filament. This causes a reduction in light intensity and a drop in color temperature. When a lamp has become visibly darkened, it should no longer be used for photomicrography.

This problem is effectively eliminated with the tungsten-halogen lamp. Although the lamp has a tungsten coil-filament, a halogen gas contained within the envelope inhibits deposition of tungsten on the glass or quartz envelope. The lamp retains its initial brightness and color temperature (3200 K) throughout its life. Near the end of lamp life a slight darkening and drop of color temperature *may* be observed. Tungsten-halogen lamps are available in housings from many microscope manufacturers. Microscopes with built-in tungsten-halogen lamp illumination are now common.

The tungsten-halogen lamp—available at present in a 12-volt, 60- or 100-watt operating values—is an excellent light source for photomicrography. It emits efficient high-intensity illumination from a small and compact coil filament. Lamp life is nominally 50 hours. Be careful to replace a tungsten-halogen lamp with one of equivalent color

---

*Note: Overvolting materially shortens the life of the lamp. Use this position sparingly.

temperature. (Some tungsten-halogen lamps have a color temperature of 3400 K. These generally have a shorter life and require different light-balancing filtration.) In replacing any microscope lamp—but particularly a tungsten-halogen lamp—be careful not to handle the envelope directly because fingerprints left on the lamp become burned into the glass, resulting in premature failure.

The xenon arc is another light source for photomicrography. This lamp produces illumination of high intensity and of daylight quality. The latter feature is especially important because it allows the use of daylight-type color films with little or no filtering. The arc is produced across tungsten electrodes in a clear envelope that contains highly pressurized xenon gas. The emission of the xenon arc is continuous not only in the visible spectrum but also in the long-wave ultraviolet and infrared spectral regions. Xenon-arc lamps operate from a special power supply. Care must be used in handling high-pressure lamps; safety eyeglasses are essential.

The zirconium arc is another excellent light source. It is very small—almost a point source. Its color temperature is 3200 K. The light intensity of a zirconium arc is not as high as that of tungsten-halogen or the zenon-arc lamps, but is high enough for efficient photomicrography. The 100-watt lamp is usually suggested, although lower wattages are also available. Like the xenon arc, it operates from its own special power supply.

Electronic flash is also used in photomicrography, specifically for photographing moving organisms. The illumination is of daylight quality. Since the flash is instantaneous, a continuously burning auxiliary tungsten lamp of low brightness is necessary for alignment, producing Köhler illumination, focusing, composing the specimen image, and general viewing of the specimen prior to photography. The flash should be synchronized with the operation of the camera shutter.

Mercury-vapor lamps serve as excellent monochromatic-light sources. With appropriate filters, the mercury green line at 546 nm, the blue line at 436 nm, or the 365 nm line in the ultraviolet region can be used for monochromatic black-and-white photomicrography. A mercury-vapor lamp should not be used in color photomicrography with a brightfield microscope. Since its illumination is deficient in many wavelengths, particularly red, it will not give a true rendition of the subject.

The sources listed above are those most commonly met with for general photomicrography. There are, additionally, many kinds of specialized light sources. One very important

consideration in selecting a light source is whether motion-picture or cinemicrography is contemplated. For cine work with cameras whose framing rate can be varied (i.e., those with built-in motors for slow-motion or high-speed photography), there may be a problem with alternating-current light sources. Illumination may be uneven or absent on successive frames. For this work, a lamp with a direct-current source such as the xenon arc should be selected.

Illuminators

Because complete control of illumination is necessary, photomicrographic illuminators, whether built-in or separate, should contain both a lens to project an image of the lamp filament and a diaphragm to control the size of the illuminated field in the microscope. The lens is usually called a *lamp condenser* or *collector lens*. Other names for the lens are field condenser and lamp collector. The term lamp condenser is used to distinguish this condenser from the microscope substage condenser, which is also used in setting up correct illumination. The diaphragm in the lamp is a variable iris diaphragm and is called a *field diaphragm*, because it controls the size

of the illuminated field of view in the microscope. Other names for this diaphragm are lamp diaphragm and radiant field stop. The terms field diaphragm and lamp diaphragm are used to distinguish this diaphragm from the microscope aperture diaphragm located in the substage condenser assembly.

Another refinement, not always included on an illuminator, is a facility for centering the light source with respect to the lamp condenser lens. We often assume that a light source in a lamphouse is centered, but this is not always true. Lamps themselves vary in regard to the position of the filament in the envelope. A centering facility corrects any discrepancy. Some microscopes have "precentered" lamps; if these prove to be uncentered, they can only be discarded.

Since filters are almost always used in photomicrography, a filter holder should also be included on the illuminator. Filters are often placed in front of the field diaphragm, but care must be taken here to make sure that any filter placed here is scrupulously clean.

If the filter holder is too close to the field diaphragm any dirt on the filter will be imaged slightly out of focus in the plane of the specimen and will be photographed along with the specimen. In actual fact, the best location for filters is in a filter holder or carrier below the substage condenser. This

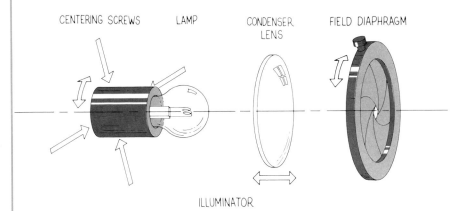

CENTERING SCREWS   LAMP   CONDENSER LENS   FIELD DIAPHRAGM

ILLUMINATOR

Fig. 3–1
MICROSCOPE ILLUMINATOR—*The essential elements of the illuminator are the lamp, a condenser lens, and a lamp, or field, diaphragm. The lamp usually can be moved laterally in any direction or rotated about its axis. The condenser lens can be focused, and the diaphragm is adjustable.*

will eliminate the possibility of out-of-focus dirt in the final image.

Some illuminators have provision for placing filters near the filament or in front of the lamp condenser. This is all right for glass filters, but does not allow the versatility of using gelatin filters because of heat from the lamp. The placement of filters will be discussed in greater detail in the section on filters.

Familiarization with the Illuminator

Separate and built-in illuminators have somewhat different means of focus and alignment. Separate illuminators should have a focusing condenser lens or lens system. Focusing is usually accomplished by a knurled focusing knob at the side of the illuminator or by a knurled ring surrounding the illuminator. In either case, adjustment of the control changes the position of the filament with respect to the lamp condensing lens. In a few illuminators the lamp condenser lens is fixed and the lamp-bulb holder is movable, the lamp or the lens, as long as the distance between the two can be varied.

Lamp focus is best understood if the illuminator is turned on and a piece of paper is held anywhere in front of the lamp, say at 0.5 to 1 metre. Turn the lamp condenser focusing knob until the filament is in focus (or slide the lamp-bulb holder back and forth in its housing). Note the size and intensity of the projected image. See how close you can bring the screen (the piece of paper) to the lamp and still focus the filament image. You will find this distance to be about 6 to 12 centimetres. The filament image will be relatively small (8 to 10 mm on a side for a flat-coiled filament) and very intense.

Now direct the filament image onto the wall at the opposite end of the room. It will not be in focus. Adjust the lamp condenser focusing knob until the filament image on the distant wall is in sharp focus. (There will be less glare if the field diaphragm is stopped down slightly.) Note that the image is now much larger (about 1 meter on a side), but much reduced in intensity. You could do the same thing projecting the filament image onto a building across the street, for example, where it would be still larger and dimmer. Thus you can see that you have full and absolute control over the location and size of the filament image from very small, very bright, and relatively close to the illuminator to very large, very dim, and distant from the illuminator—all by adjusting the lamp condenser focus (or filament-condenser lens distance).

But of all the possible places you could locate the filament image, it belongs in one and only one place, viz, the plane of the microscope aperture (substage) diaphragm.

As for the size of the filament image, it must be just large enough to completely illuminate the aperture diaphragm, that is, the entrance pupil of the microscope substage condenser. If the image is too small—say, one-half the diameter of the aperture diaphragm—the microscope condenser will not be able to illuminate the specimen fully for high numerical aperture objectives. This condition occurs if the separate illuminator is placed too close to the microscope. If the filament image is too large, as when the illuminator is too far from the microscope, the filament image outside the condenser aperture diaphragm is not utilized and light is lost unnecessarily.

A good separate illuminator will also have some means of aligning the filament to the optical axis of the condensing lens. This is necessary because, with the exception of special lamps that are made to have their filament in a specific place within the envelope, the filament location varies from lamp to lamp. For alignment of the lamp filament with the optical axis of the illuminator, centering screws are necessary. Additionally, the lamp bulb socket should be rotatable in its mount and provided with a lock screw to hold the socket in any position.

To align a lamp filament to the optical axis of the illuminator, turn the lamp on and project a focused image of the filament on a distant wall. (Projecting on a distant wall will magnify any errors in the centering.) Now, loosen the lamp-socket lock screw slightly and rotate the entire lamp socket in its mount. Note what the filament image does when the socket is rotated.

Ideally, the filament image should rotate in one spot about its axis. If on rotating the socket the filament image moves in an arc, the filament is off-axis and it must be brought on-axis by adjusting the two mutually perpendicular or coaxial centering screws. Note where the center of rotation of the arc should be and bring the filament image to that point with the centering screws. Check the alignment by again rotating the entire socket in its mount. Continue adjustment of the centering screws until the filament image rotates about its own axis. Then turn the lock screw to secure the entire socket in its holder. The filament is now aligned and on-axis with the lamp condenser. This alignment needs to be repeated only when the lamp is replaced.

The separate illuminator must also have an adjustable iris diaphragm (the field diaphragm). Ideally, the adjustable field

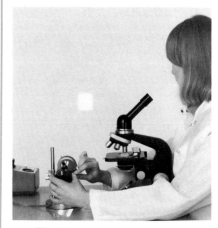

Fig. 3-2
CENTERING LAMP—*Determine if the lamp filament is centered by rotating the lamp socket while observing the projected image of the filament. The image should rotate about its own axis and not move in an arc.*

diaphragm should not change position when the lamp condenser is adjusted; the condenser lens should move internally or the filament should move with respect to the fixed lamp condenser lens. The field diaphragm will be focused in the plane of the specimen by means of the microscope substage condenser. If the field diaphragm moves every time the lamp condenser is adjusted, the microscope substage condenser will have to be readjusted.

Separate illuminators should also be on adjustable stands so that their height and angle of tilt are fully under the photomicrographer's control.

Amongst the best lamps for separate illuminators are those with flat-coil filaments. Next best are those with tightly wound helical coils. The worst filaments are the loose coils that are themselves coiled ("coiled coils"). These have too much depth to be useful without a ground glass in the illuminator beam—which you decidedly do not want because you cannot then focus the filament.

In summary, the ideal separate illuminator has

- A focusing condenser system,
- An adjustable field diaphragm that does not change position when the condenser is focused,
- A mount to rotate the lamp socket, with a locking screw or clamp,
- Lamp centering screws,
- A tight, flat coil filament,
- Height and tilt adjustments.

Built-in illuminators may have adjustable lamp condenser focus by means of a focusing lever that protrudes from the illuminator. More commonly, the lamp condenser lens is fixed in the base and focusing is accomplished by moving the lamp socket in and out, i.e., closer to and away from the fixed lamp condenser lens. Built-in illuminators may also be supplied with filament centering screws. If they are not provided, centering of the filament is accomplished by rotating the entire lamp socket in its mount until the filament is centered and then locking it in place with the lock ring or screw.

Microscopes with built-in illuminators have their field diaphragms located in one of two places, either vertically somewhere in the base between the lamp and the 45° reflecting mirror or, more commonly, horizontally in the base just below the microscope condenser. Alignment of the filament to the optical axis of built-in illuminators is accomplished with the aid of a piece of ground glass or translucent plastic or a piece of paper placed on top of the field diaphragm in the microscope base. The filament can be *temporarily* brought to focus in the plane of the glass or paper by moving the lamp in or out and then centering the filament image with respect to the field diaphragm, using the lamp centering screws. Don't forget to refocus the filament image in the plane of the aperture diaphragm.

Mastering all the adjustments of the illuminator and knowing the reasons for these adjustments is necessary for a thorough understanding of subsequent microscope illumination methods. If you do not understand the parts and functions of the microscope illuminator, reread this section before proceeding.

Methods of Illumination

One can divide the methods of illumination on the basis of specimen transparency or opacity. If the specimen is very thick or even opaque, reflected-light methods (incident light, top light, epi illumination) must be used. Metallography utilizes the reflected-light method of illumination and is discussed later in this book. The majority of subjects encountered in general biomedical and industrial photomicrography are transparent or can be made transparent and are examined and photographed by transmitted illumination.

In basic terms, this involves directing the light from the lamp filament to a plane mirror, through the substage condenser, through the subject and the field around it, and into the microscope objective. If the illumination system has not been properly adjusted, a photomicrograph is likely to be disappointing. This is true even if the highest quality optics have been used, the image has been focused critically, and the exposure of the film has been correctly determined. The two essential requirements are (1) that the whole illumination system be centered, and (2) that the cone of light from the illumination system completely fills the aperture of the microscope objective (see Fig. 3-3).

Historically, the first illumination methods utilized daylight, candlelight, or the light from oil lamps, with or without a bull's-eye condenser lens to concentrate the light on the specimen. By the days of "brass and glass" (latter half of the 19th century) the photomicrographer's favorite source of illumination was a white cloud in the northern sky. Needless to say a white cloud in the northern sky was not always available when the photomicrographer was ready to make a photograph.

Nelsonian (Critical) Illumination

Based upon principles demonstrated by Ernst Abbe, the noted English microscopist Edward Nelson devised a splendid form of illumination that was used successfully for many years. Nelson's method consisted of focusing the side of the oil lamp flame in the plane of the specimen to achieve the much desired self-luminous specimen condition. The light was pleasant and even because of the homogeneity of the flame. This method of illumination—usually called *critical illumination* but more properly termed *Nelsonian* (or sometimes *Abbe-Nelsonian*) *illumination*—is seldom used today. The reason is that today's electric light sources use coils of tungsten wire. If you were to focus the source in the plane of the specimen you would see and photograph the filament coils superimposed on the specimen—not a very pleasant sight. This is a result of the nonhomogeneity of a wire coil. The closest source to Nelson's homogeneous wick available today is the ribbon filament lamp.

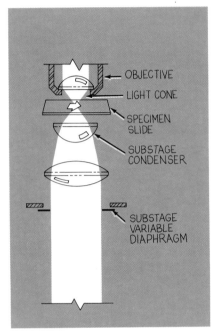

Fig. 3-3
CONE OF ILLUMINATION—
The substage condenser must be focused and the diaphragm adjusted so that the cone of illumination completely fills the sperture of the microscope objective.

Here instead of a coil of tungsten wire, a flat, ribbon-like piece of tungsten metal forms a nearly homogeneous source. Illuminators that use this 6-volt, 18 amp lamp can be used to set up Nelsonian illumination because the flat ribbon provides a uniformly lighted field of view even when focused in the plan of the specimen.

Actually, there are at least two instances when the microscopist may wish to set up Nelsonian illumination even with nonhomogeneous sources. One is for darkfield work and the other is for fluorescence microscopy of specimens with very weak primary fluorescence.

The precise steps for setting up this method of illumination will be obvious after a discussion of *Köhler illumination* and will involve only a different adjustment of the lamp condenser. (Remember from the previous discussion of **Familiarization with the Illuminator** that you have complete control over the location of the filament image by adjustment of the lamp condenser.)

Köhler Illumination

Köhler illumination is the most common system of transilluminating a microscope specimen in photomicrography. It is the method that results in the most successful microscopy and photomicrography. It is used for visual work as well as photomicrography because it provides the *highest intensity* of *even illumination* from *nonhomogeneous sources*. This section on optimum illumination—like lighting in all phases of photography—is the *most critical* aspect of good photomicrography and it must be learned and practiced until mastered.

Köhler illumination does not inherently provide more intense illumination than the Nelsonian system of focusing the light source on the specimen; often the reverse is true. In Köhler illumination, a central fraction of the lamp condenser (a small relative aperture) may be focused in the plane of the specimen. For fluorescence work, Nelsonian illumination may be used to provide more illumination in the field.

August Köhler introduced his method of illumination in 1893. Köhler devised his system to obtain uniform illumination from nonhomogeneous sources such as the newly introduced electric lamps. It still serves this purpose today. Köhler, himself, worked with an optical bench setup. Modern application of Köhler's principles in integrated microscope systems has become complex and even, in some cases, compromised.

In the simplest, original arrangement, the light source (lamp filament) is focused in the

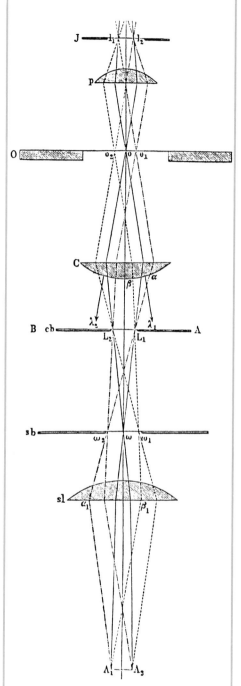

Fig. 3-4
KÖHLER ILLUMINATION—
This basic arrangement of microscope elements provides uniform illumination from nonhomogeneous sources. The illustration here is adapted from an original drawing in Dr. August Köhler's paper, "Ein neues Beleuchtungsverfahren für mikrophotographische Zwecke," published in 1893.

rear focal plane (entrance pupil) of the substage condenser. With a given lamp condenser and source size, the distance of the lamp is such that the image of the source just fills the aperture of the substage condenser in use. The required distance changes with magnification and some other variables. In practice, since the setup (lamp distance) is rarely changed, the image of the source is usually kept large enough to fill the whole aperture of the condenser. This is photometrically inefficient for lower magnifications.

The substage condenser is focused to bring the image of the lamp condenser into the plane of the specimen. This is because the aperture plane of the lamp is uniformly illuminated by any source that is neither very large nor has a variably shadowed shape (biplane filaments). Putting a lens in the aperture does not change this. Ideally the field diaphragm is against the lamp condenser.

Setting Up Köhler Illumination

Basically, the principles of Köhler illumination applied to the modern microscope can be reduced to (1) focusing the image of the field (lamp) diaphragm in the plane of the specimen using the microscope substage condenser, and (2) focusing the image of the filament in the plane of the aperture (substage) diaphragm using the lamp condenser (collector lens). Refer to Figures 3-5 and 3-6 throughout the following discussion.

While two separate ray paths, the image-forming beam and the illuminating beam, are shown on individual diagrams (Fig. 3-5 and 3-6), they do not have any actual separate existence. The separation is simply a graphic device to assist visualization of the microscope setup. The diagram showing the illuminating-ray path (Fig. 3-5) demonstrates that each point of the lamp filament is in focus at the plane of the aperture diaphragm and at the objective back focal plane. Furthermore, each point of the lamp filament is shown to contribute to full illumination of the specimen and hence to the subsequent film image. Conversely, the image-forming ray diagram (Fig. 3-6) shows that the lamp filament fully illuminates the field diaphragm which in turn is imaged in the plane of the specimen. The uniformly illuminated specimen is focused by the objective at the intermediate image plane. This real image is further magnified by the eyepiece which renders the rays parallel for subsequent imaging by the infinity-focused lens of the eye or of a camera.

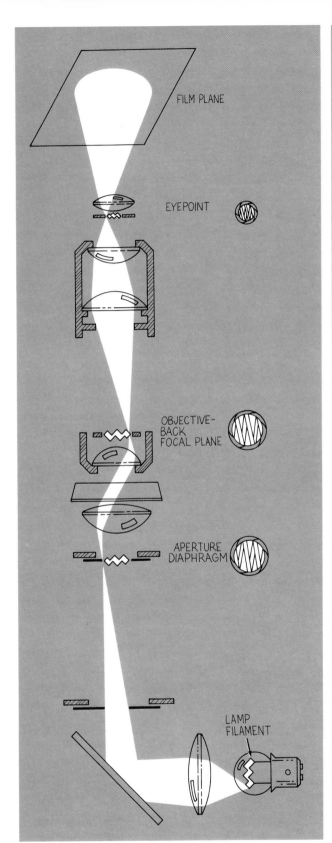

Fig. 3-5
ILLUMINATING RAY PATHS
*are traced from one end of the lamp filament. Conjugate foci of lamp filament are aperture diaphragm, objective back focal plane, and eye-point. At the right, is the image that would be seen at each conjugate.**

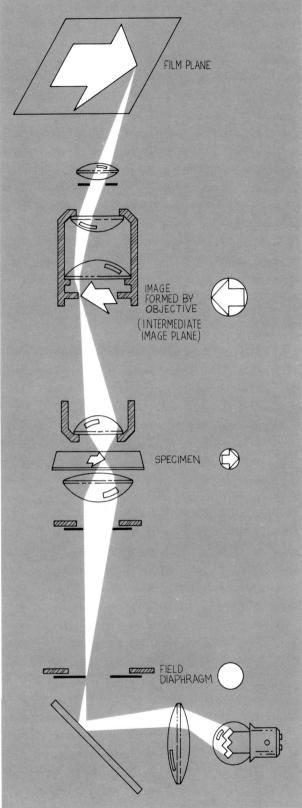

Fig. 3-6
IMAGE-FORMING RAY PATHS *are traced from two ends of lamp filament. Conjugate foci are field diaphragm, specimen plane, intermediate image plane (entrance pupil of eyepiece), and, with camera in place, the film plane. See text at left for further explanation.**

Note: While ray traces are schematically correct, optical components are represented by simple lenses and element size and spatial relationship are not to scale.

Center the Lamp Filament

To set up Köhler illumination, open all diaphragms and follow these steps:

(Using separate lamp; start here.)

1. If you are using a separate lamp, i.e., one that is not built in, start by centering the filament to the optical axis of the lamp condenser as previously described under **Familiarization With the Illuminator**. To repeat, briefly, you do this by turning the lamp on and projecting an image of the filament on a distant wall or other plane surface. Loosen the bulb lock screw, rotate the entire bulb holder, and observe the projected image. If the image of the filament does not rotate about its own axis, adjust the bulb centering screws until it does, and lock the rear assembly in place with the thumbscrew lock. This step will not be necessary if your lamp is factory precentered. After ensuring that the filament is on the optical axis of the lamp condenser, place the illuminator low on its stand and initially about 15 cm in front of the microscope mirror. Use the plane (flat) side of the mirror. Direct the light to the center of the mirror (this can be checked by temporarily placing a piece of paper, card, or ground glass on the mirror, Fig. 3-7). Remove the paper or ground glass and tilt the mirror so that the light is directed up into the condenser, Fig. 3-8. (Hint: If the microscope slide is moved until the label is over the stage opening, the spot of light projected on it will indicate the best position for the mirror.)

(Using built-in illumination; start here.)

1a. If the microscope has built-in illumination, center the filament by placing a piece of paper, translucent plastic, frosted glass, or an accessory centering aid on top of the opening in the stand where the light comes out, Fig. 3-9 (on some microscopes, this may also be where the field diaphragm is located). If the opening is not evenly illuminated, use the centering screws or rotate the lamp in its holder to obtain a centered light source, Fig. 3-10.

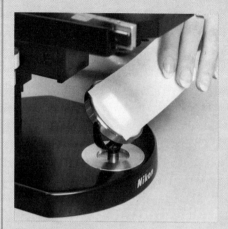

Fig. 3-7

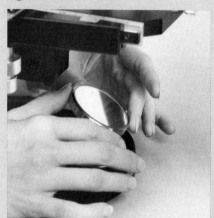

Fig. 3-8

Fig. 3-10

Fig. 3-9

Focus the Specimen

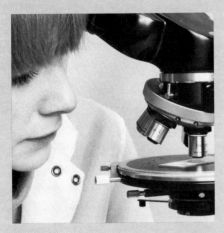

Fig. 3-11

2. Move the specimen on the stage into the light path and lower the objective (10X to start with), while watching from the side, until it is close to the cover glass, Fig. 3-11; and then while looking into the eyepieces, focus *up* (moving the objective and specimen apart) until the specimen is in focus. It is bad practice to focus downward (moving objective and specimen together) while looking into the microscope eyepieces because the specimen plane is easily overshot, resulting in possible damage to both the objective and the specimen. If you miss the correct focus, start over by lowering the objective while watching the operation from the side, and again focusing upward. (Hint: If the microscope is equipped with a rotating stage, use one hand to rotate the stage back and forth in a short arc while the other hand focuses up slowly with the coarse focus adjustment. It is easier to spot the correct plane of focus when the target has some motion across the field of view. A built-in or attachable mechanical stage can be used in the same way, i.e., making the specimen move left to right or up and down with one hand while the other hand adjusts the coarse focus.)

3. With the specimen in focus and illumination adequate, if as yet unrefined, center objectives that are in centerable nosepiece mounts. This step is for microscopes with rotating stages; microscopes with fixed stages do not require the adjustment.

4. If you are using a binocular or trinocular head, adjust the binocular tubes for interpupillary separation and diopter correction, Fig. 3-12. First adjust the separation between the two tubes (as you would a pair of binoculars) until the separation between them matches your interpupillary distance, which is indicated on the dial or scale, so that you can reset it if more than one person uses the microscope. (The average interpupillary distance is 65 mm.)

Now, if one of the eyepieces has a scale or cross hairs or other graticule, rotate the *eye lens* of that eyepiece (take care that the whole eyepiece does not turn) until the scale is in perfect focus, Fig. 3-13.

Note whether your binocular head has one or two diopter adjustment collars—most have one. The diopter adjustment collar is used to make up for differences between your two eyes. If your binocular head has one diopter adjustment collar, proceed as follows. Use a file card or something similar to cover the eyepiece that has the diopter adjustment. (Do not just shut the one eye; strain will eventually result.) Look for some tiny detail or speck of dust in the specimen plane through the other eyepiece and focus on it carefully with the microscope fine adjustment. Now cover the eye you have just been using and focus on the same fine detail or speck of dust with the other eye by using the diopter adjustment collar, Fig. 3-14. *Do not use the fine adjustment.* Remove the card and the binocular head will be adjusted correctly for both eyes.

Incidentally, the microscope will make up for farsightedness and nearsightedness, so that eyeglasses for these conditions do not have to be worn when using the microscope. The microscope cannot be adjusted for astigmatism, and if the astigmatism is severe, corrective eyeglasses must be worn when using the microscope.

Fig. 3-12

Fig. 3-13

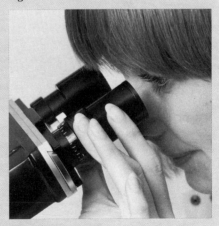

Fig. 3-14

Fig. 3–15

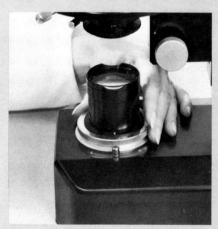

Fig. 3–16

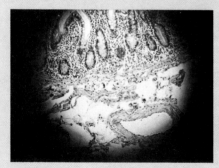

Fig. 3–17

Adjust the Field Diaphragm

5. With the objectives centered and the binocular head properly adjusted for both eyes, you are ready to start the most critical steps in achieving Köhler illumination.

Close the field (lamp) diaphragm, Fig. 3-15 or Fig. 3-16, until you can see all of it within the field of view, Fig. 3-17. It is likely that the image of this diaphragm will be neither centered nor well defined. Tilt the mirror until the out-of-focus image of the field diaphragm is centered in the field of view. With built-in illuminators, adjust the substage condenser centering screws to get a centered image, or adjust the sliding centering device in the base—whichever is provided.

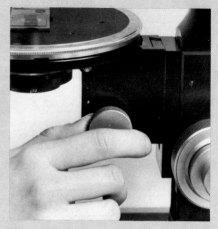

Fig. 3–18

Fig. 3–19

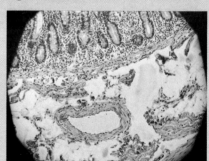

Fig. 3–20

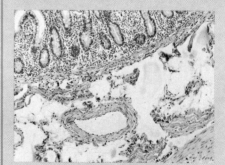

Fig. 3–21

Focus the Field Diaphragm

6. Focus the now-centered field (lamp) diaphragm by adjusting the substage condenser up or down. This can be done with the rack-and-pinion focus adjustment, Fig. 3-18, or by loosening the lock screw and sliding the condenser up or down by hand. After the field diaphragm is brought into sharp focus, Fig. 3-19, open the diaphragm until it is just outside the field of view, Fig. 3-20 and Fig. 3-21. (Lock the condenser in place if a lock is provided for the focusing mechanism.)

Focus the Lamp Filament

7. Focus the lamp filament in the plane of the substage diaphragm by using the lamp condenser focus adjustment, and center the filament image, Fig. 3-22 and Fig. 3-23. This step will be impossible when substage components intervene or if the diaphragm is between the condenser lenses. However, since the specimen is focused and the substage condenser is focused to image the field diaphragm, there is a conjugate focal plane of the substage diaphragm that can be used instead. This is located at the objective back focal plane. This is most easily seen by simply taking the eyepiece out and looking down the tube toward the back of the objective. (Alternatively, one can insert the Bertrand lens of a polarizing microscope, insert a phase telescope in place of the eyepiece, employ a built-in phase telescope, insert a pinhole eyecap in place of the eyepiece, use a Klein lupe over the eyepiece, or view the exit pupil of the eyepiece with a magnifier.) At the objective back focal plane, you *should* see the filament and the aperture (substage) diaphragm, but initially you probably will not. Open the substage aperture diaphragm if you have not already done this. If the filament is not in focus when you look down into the body tube at the back focal plane, Fig. 3-25, adjust the lamp condenser focus (on a separate illuminator) until the filament is in focus at the back focal plane, Fig. 3-26. If the filament is not centered, tilt the entire lamp assembly up or down or rotate it about a vertical axis until the filament is centered, Fig. 3-27 and Fig. 3-28. *Do not touch the filament centering screws; these were used to align the filament to the optical axis of the illuminator.*

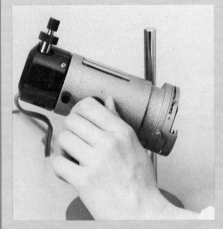

Fig. 3-22

Fig. 3-23

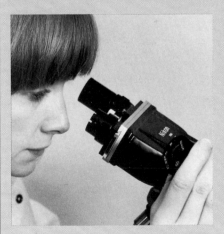

Fig. 3-24

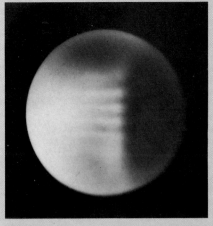

Fig. 3-25

Fig. 3-26

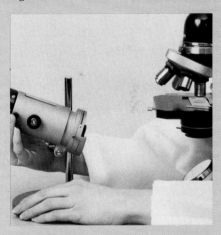

Fig. 3-27

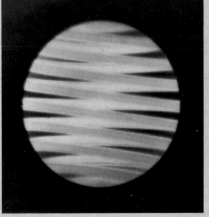

Fig. 3-28

If the illumination is built in, loosen the lamp lock screw and move the bulb in and out of its holder and, at the same time, rotate it until a position is found where the filament is centered and focused. Lock the bulb in place. Many microscopes made today with built-in illumination incorporate somewhere in the light path a diffusing element of either frosted or opalized glass or with an *orange peel* or *lemon peel* surface. These may be separate pieces of glass or the peel may be chemically etched directly to one surface of the lamp condenser lens. In any case, a sharp image of the lamp filament will not be possible. In this event, while looking down the body tube at the objective back focal plane, move the bulb back and forth in its socket or rotate it and watch for the most intense, even illumination of the back focal plane, Fig. 3-29 and Fig. 3-30. Lock the bulb in place. Focusing of the lamp filament may also be necessary to eliminate bright or dark spots.

There may be some question as to whether Köhler illumination is possible if a diffuser is in the system. Externally mounted light sources of the better manufacturers are interfaced to the microscope via a connecting tube containing relay optics. These relay optics form an intermediate image of the filament exactly at the location of a low scattering diffusion disk. The diffuser thus becomes the source and can be treated just as a filament would. In this way, precise Köhler conditions are maintained. But indiscriminate placement of a diffuser makes strict Köhler conditions impossible. To repeat, if a diffuser is in the system, a sharp image of the filament will not be seen, but you can adjust the lamp focus for a fully and evenly illuminated back focal plane. Some microscopes illuminators are provided with a swing-in/swing-out diffuser so the photomicrographer has a choice; in such cases, the diffuser is used for low-power work only.

Adjust the Aperture Diaphragm

8. Now, after the objective back focal plane is fully illuminated with a centered, focused filament image, close down the substage aperture diaphragm while viewing the back focal plane, until the diaphragm just comes into the field of view, Fig. 3-31 and Fig. 3-32. The image of this diaphragm may not be sharp for two reasons. (1) The specimen itself may be diffusing the filament image. Move the

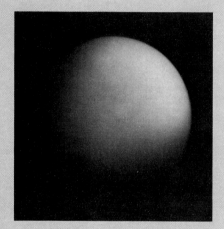

Fig. 3–29

Fig. 3–30

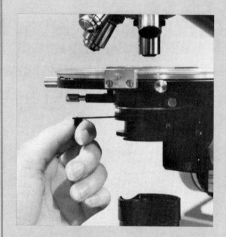

Fig. 3–31

Fig. 3–32

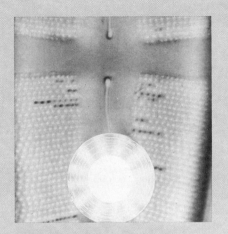

Fig. 3–33a

Fig. 3–33b

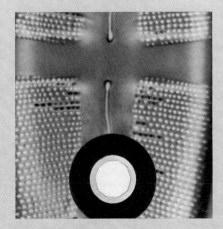

Fig. 3–33c

microslide to a location where the specimen is thin or at the edge of the field of view. (2) If a Bertrand lens or phase telescope is used and it is not focusable, the image of the diaphragm will not be sharp. The use of a pinhole cap will sharpen up the image. If a commercial pinhole eyecap is not available, simply punch a hole in a piece of paper with a dissecting needle and place the paper with the pinhole directly on the body tube with the eyepiece removed. Place your eye very close to the hole for viewing the back focal plane.

You may also notice at this time that the aperture diaphragm is not exactly centered with respect to the back focal plane. On early microscopes, the aperture diaphragm is rotatable and adjustable sideways via a rack-and-pinion gear, and exact centering is very easy to accomplish. In the absence of this feature, one only has recourse to appeal to the manufacturer for more precise centration. (Generally there are three tiny screws holding the aperture diaphragm onto the condenser. Adjusting these three screws will center the diaphragm. This need only be done once.)

The adjustment of the aperture diaphragm is of the utmost importance in careful microscopy as it controls the numerical aperture of the system and therefore the ultimate resolving power, depth of field, and character of the image. The exact working aperture of this diaphragm is frequently recommended in restrictive terms, such as two-thirds, four-fifths, or nine-tenths of the diameter of the fully illuminated back focal plane.* Actually, the setting should vary with the correction of the objective and the nature of the specimen. Remember that the specimen becomes part of, and alters, the optical system of the microscope. The aperture diaphragm must be adjusted differently to accommodate the optical characteristics of each inividual specimen. The photomicrographer must critically evaluate the specimen image while adjusting the aperture diaphragm until the optimum compromise position is found between depth of field and resolution. There will be one best position. (One well-known microscopist used to advise starting with a wide open diaphragm and closing down slowly until the image suddenly acquired "oomph.") Leave the diaphragm open too much and the image lacks contrast and depth of field, Fig. 3-33a; close it down too much and diffraction lines surround the now too-contrasty

*See **Diffraction Theory and Resolution**, page 9.

image, Fig. 3-33b. The best compromise setting must be found, Fig. 3-33c. The tendency on the part of beginning photomicrographers is to close the diaphragm excessively. The critical photomicrographer is constantly adjusting the aperture diaphragm for optimum position. *NEVER use the aperture diaphragm merely to control brightness of the illumination.* Do this with neutral density filters for color photomicrography. For visual work and black-and-white photomicrography, you may turn down the transformer rheostat.

9. Put the eyepiece back in the body tube and observe the image. The image should now be brightly and evenly illuminated with good depth of field and resolving power. If the lamp was moved during Step 7, you will again want to close the field diaphragm until it comes into the field of view to check its centration. If it is not centered, tilt the mirror or adjust the condenser centering screws. The microscope and illumination have now been properly prepared for photomicrography.

As objectives are changed, the adjustment of the two diaphragms must be altered. If the objective NA is increased, which generally happens when magnification is increased, the light-gathering ability of the objective is increased. The objective requires a larger illuminating cone, so the substage aperture diaphragm must be *opened* to allow a wider illuminating cone of light. At the same time, the field size within the specimen plane is *decreased* because of the increase in magnification. The field diaphragm must be closed down until it is again in the field of view and then opened until it is just outside. The operation and effect of the adjustment of both diaphragms are best seen in practice by actually changing objectives (such as 10X to 40X).

In photomicrography, the field diaphragm is adjusted until the diameter of its image is equal to, or just larger than, the diagonal of the film format. The image can be seen and adjusted in size while looking either through a side telescope or at the film plane of the camera (as on a ground glass). But be careful when making this adjustment with reference to the film format graticule as seen in the side telescope of photomicrographic cameras because these format indicators are frequently undersized. Initially it is better to open the photomicrographic camera back, place a ground glass at the film plane, and locate an image of the field diaphragm by closing it down until it just enters the field of view. Then look through the side telescope to see where the image of the field diaphragm lies with respect to the format-indicating lines.

Test Yourself on Köhler Illumination

Setting up Köhler illumination is the basis of careful photomicrography. To reinforce the steps taken in setting up this kind of illumination and the reasons for each, take this quiz, performing each demonstration in turn for yourself.

Q. Where does the *first* real image of the field (lamp) diaphragm appear?

A. In the plane of the specimen (see Fig. 3-5 and 3-6.).

Q. How do you get the field diaphragm in focus in the plane of the specimen?

A. By adjusting the position of the substage condenser (see Step 6).

Q. How can you show that the field diaphragm is really in the plane of the specimen?

A. Place a piece of paper directly on top of the specimen slide and observe the disk of light (*not* by looking through the microscope, but by looking from the side at the paper directly, Fig. 3-34). Open and close the field diaphragm and observe how the diameter of the spot of light changes.

(Remove the piece of paper from on top of the specimen microslide.)

Q. Where is the second place the real image of the field diaphragm occurs?

A. At the intermediate image plane, along with the specimen (see Fig. 3-5 and 3-6).

Q. Where is this intermediate image plane located, and how can you show that the specimen and field diaphragm image are located there?

A. The intermediate image plane is located about one centimeter (about ½ inch) below the top of the tube where the eyepiece is fitted. Remove the eyepiece and place the piece of paper over the opening of the body tube (or, more properly, bend the end of a narrow strip of paper into an L-shape and introduce it down the body tube about one centimeter, Fig. 3-35). You will see an image of the specimen on the piece of paper (specimen focus may have to be changed slightly and the light intensity increased somewhat). Again open and close the field diaphragm. The field diaphragm will be seen to open and close right over the specimen image.

Q. Good. Now where is the third place you can locate the image of the field diaphragm and specimen?

A. On the retina if you are using the microscope visually, or on the film plane if you are making a photomicrograph.

Q. How can you show this?

A. Replace the eyepiece and hold a piece of paper about 15 centimeters (about 6 inches) above the eyepiece (it helps to subdue the room lights). You will see the specimen projected on the paper, Fig. 3-36 (touch up the focus and intensity if necessary). As before, open and close the field diaphragm. You have just traced the conjugate foci of the field diaphragm and demonstrated each position as shown in Fig. 3-5 and 3-6.

Q. If the field diaphragm had not been in sharp focus at the intermediate image plane or at the projected plane above the eyepiece, what would this indicate?

A. An out-of-focus field diaphragm anywhere indicates that the substage condenser focus is not correct. (Remember, it is always the substage condenser that controls the focus of the field diaphragm.)

Q. Where does the first real image of the filament occur?

A. The first real image of the filament should appear in the plane of the aperture (substage) diaphragm (see Fig. 3-5 and 3-6).

Q. How can you show that the filament image is at the aperture diaphragm?

A. With some microscopes, just look up under the condenser. If the aperture diaphragm is located outside the condenser system, the filament can be seen projected on it. If the diaphragm is between the lenses, place a piece of paper as high up under the condenser as possible and the filament will be seen on it slightly out of focus (it's in sharp focus in the actual plane of the diaphragm). If the microscope has a separate illuminator, one can frequently just look at the reflection in the substage mirror to see the filament on the leaves of the aperture diaphragm, Fig. 3-37.

Q. If the filament is not in sharp focus in the plane of the aperture diaphragm, what is out of adjustment?

A. The lamp condenser (collector) is not focused properly (see Step 7).

Q. Where is the second plane (conjugate focus) after the aperture diaphragm that you see the image of the filament?

A. In the objective back focal plane (see Fig. 3-5 and 3-6).

Q. How can you show that the filament is focused in the objective back focal plane?

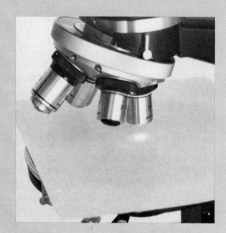

Fig. 3–34

Fig. 3–35

Fig. 3–36

Fig. 3–37

Fig. 3–38

A. The simplest way is to remove the eyepiece and look down the tube. Do this. It may help to make a crude pinhole eye cap. With the eyepiece removed, place the piece of paper with a pinhole right down on the top of the opening in the body tube. Place your eye close to the pinhole and you will get a very clear view of the objective back focal plane (you may have to turn the light intensity down).

Q. While you are viewing the objective back focal plane, what else should you see here besides the filament image?

A. The image of the aperture (substage) diaphragm. While viewing the filament at the objective back focal plane with the pinhole eye cap or accessory telescope, open and close the aperture diaphragm. You will see it open and close over the filament image. Both of these images may not be sharp if there is too much specimen in the field of view. Remember, the filament and aperture diaphragm images must pass through the specimen plane to get to the objective back focal plane.

Q. Where is the next location (conjugate focus) of the lamp filament image (and aperture diaphragm)?

A. At the Ramsden disk—the eyepoint (see Fig. 3-5 and 3-6).

Q. How can you show that the filament and aperture diaphragm images are located at the eyepoint?

A. First, locate the eyepoint. Hold a piece of paper directly on top of the eyepiece. You will see a round spot of light. *Slowly* raise the paper and observe the diameter of the spot of light. The spot of light at first gets smaller in diameter and then, after reaching a minimum, it gets larger. Hold the paper at the *minimum* diameter, Fig. 3-38. That is the Ramsden disk (the eyepoint, the exit pupil). It is this tiny spot where the lens of your eye is placed when the microscope is used visually, or where the shutter, auxiliary projection lens, or beam splitter goes when a photomicrograph is made. The distance of this eyepoint above the top of the eyepiece is known as the eye *relief*. It varies from just a couple of millimeters to about 25 mm for high-eyepoint eyepieces. Now, while holding the piece of paper carefully at the Ramsden disk with one hand, open and close the aperture (substage) diaphragm with the other hand. You will see the spot diameter change. The filament image will be too small to see, but to see an enlarged image of the eyepoint, remove the paper and focus on the spot with a pocket magnifier of 3X to 10X (or use a positive eyepiece turned upside down). You will have to focus carefully and keep your head steady, but if you do this correctly you will be seeing an image of the objective back focal plane with its filament and aperture diaphragm.

Q. What is out of adjustment if, in these demonstrations, the filament is not in sharp focus?

A. If the filament is not in sharp focus, the lamp condenser (collector) is not properly adjusted (see Step 7). Remember, the lamp condenser focus setting determines the location of the filament image. You have now completed tracing the image of the filament through the illuminating ray path (Fig. 3-5).

You should repeat the steps in setting up Köhler Illumination until they become firmly fixed in your mind and become second nature. If you have followed the steps carefully, you have adjusted the microscope and the illumination for the best possible visual image. The microscope is now ready for attachment of the camera.

Difficulties in Köhler Illumination

High-Power Objectives

One difficulty in setting up Köhler illumination may occur when a high-magnification, high-NA objective is being used. The problem that may be encountered is that an image of the field diaphragm will not be obtained. This is due to the inability of some field diaphragms to close down sufficiently far or to the inability of the substage condenser to produce a sufficiently small field diaphragm image. In this case, two methods of resolving the difficulty are possible. Center the field diaphragm with a lower magnification objective first and then close the diaphragm as much as possible when the high-power objective is used. Or temporarily decenter the substage condenser (or tilt the mirror) until an edge of the field diaphragm can be seen and use this on which to base the focus adjustment of the substage condenser.

Another problem may be the inability to get a sharp image of the field diaphragm with high-numerical-aperture condensers. This will be due to the use of slides that are too thick.

Low-Power Objectives

Problems will also occur when very low magnification objectives (1X to 4X) are used. With such low-power objectives, Köhler illumination no longer applies. There are several ways of dealing with this. With some microscopes, you can remove the top lens of the condenser (by unscrewing it, by flipping it out, or by lifting it off); this has the effect of reducing the NA of the condenser and increasing the diameter of the illuminating cone. Some microscope condensers have a slide-in auxiliary lens to change the effective focal length of the condenser or a flip-in diffuser element. In still other microscopes, the entire condenser must be replaced. In the case of microscopes with separate illuminators, you can remove the condenser entirely and use the concave side of the mirror for illumination. Leave the aperture diaphragm wide open, and the field diaphragm will act as an aperture diaphragm. Consult the microscope manufacturer's recommendation regarding the use of very low magnification objectives, as this differs with the design.

Image Brightness

Very often when the microscope and the illumination are correctly adjusted, the image in the microscope will be extremely bright—too bright, in fact, for comfortable visual observation. The brightness of the image can be reduced by placing a neutral density filter in the light beam. Trial and error will indicate how dense the neutral filter should be in order to provide a comfortable level of brightness. Many built-in illumination systems incorporate neutral density filters for this purpose.

When using a separate microscope illuminator, have a neutral density filter available to reduce image brightness for visual work. In this case, place the filter in a filter holder on the front of the illuminator or under the substage condenser. Remove it when making a photomicrograph.

Although closing the aperture diaphragm decreases image brightness, image quality suffers for both visual and photographic purposes. *Never use the aperture (substage) diaphragm to control light intensity.*

Neutral filters of several densities are actually used in photomicrography to control exposure time so that it will be within the range of available shutter speeds.

A variable transformer can reduce image brightness with tungsten filament lamps. This provides a comfortable level of illumination for visual use or intensity when black-and-white films are used. Lamp life will be considerably prolonged also. For consistent color photography, use the lamp at the normal rated voltage. Tungsten-halogen lamps must not be used more than 30 percent below their rated voltage, or the tungsten recycling reaction will not occur.

Chapter Four

CAMERAS FOR PHOTOMICROGRAPHY

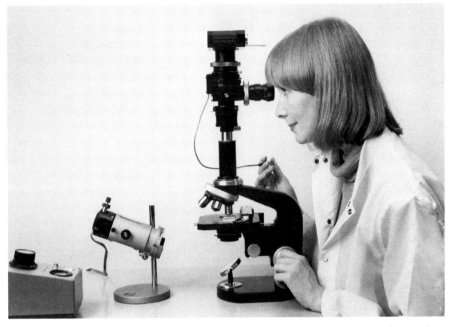

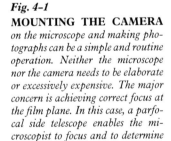

Fig. 4–1
MOUNTING THE CAMERA
on the microscope and making photographs can be a simple and routine operation. Neither the microscope nor the camera needs to be elaborate or excessively expensive. The major concern is achieving correct focus at the film plane. In this case, a parfocal side telescope enables the microscopist to focus and to determine the field to be recorded.

A correctly-prepared microscope, used to best advantage with properly controlled illumination, is the image-forming part of a photomicrographic system. The camera is the means for recording on film the image formed by the microscope. Only when the quality of the image produced in the microscope is the highest attainable can the camera record an excellent photomicrograph. If it is not there, then no film, camera, or camera refinement can improve the image quality in the photomicrograph. It is important, however, that the image *should not be degraded* in the camera or by any photographic technique.

Although almost any camera can be used to record the image, a camera specifically designed for photomicrography offers many advantages. The selection of a camera is most often governed, however, by cost and convenience.

Of course, in the strictest sense, a camera is not even needed to make a photomicrograph. A right-angle prism can be placed directly on top of the eyepiece and the image projected and focused on a nearby wall. Then, in the dark, a photographic film can be taped or otherwise attached to the wall where the projected image was focused and the microscope lamp turned on for the exposure. The obvious disadvantage to this method of making photomicrographs is the inconvenience of working in the dark for much of the time. However, it does illustrate the fact that a conventional camera is not absolutely necessary, and the method has the advantage of reducing the cost to that of the film or paper.

A step up from no camera would be to make one's own photomicrographic camera. Such homemade devices have been made from soft-drink cans, coffee cans, and, at least one highly reputable scientific work has been published whose photomicrographs were made with a shoe-box photomicrographic camera. After all, once the critical image has been formed in the microscope, it only remains to support the film in a plane upon which a focused image can be projected. Such homemade devices are fun to design and construct, and one is shown in the illustrations. Most photomicrographers will already have a conventional camera with which to make photomicrographs or a photomicrographic camera made especially for photography through the microscope.

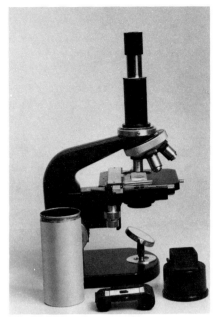

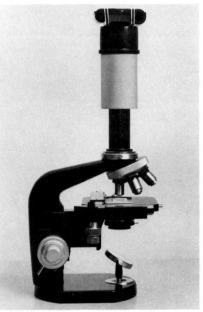

Fig. 4–2
SIMPLIST MICROSCOPE CAMERA *may be no camera at all. With a soft aluminum beverage can, the top of a shaving cream dispenser, scraps of fiberboard and aluminum, rubber bands, and a 126-size film cartridge, John Delly constructed this "film holder" to mount on the microscope tube. After emptying the can, John removed the top and made a careful cutout in the bottom of the can to fit tightly over the microscope tube. He selected the dispenser top because it fit over the can to form a lightlock and because it also was easy to cut and glue. Glued-up fiberboard pieces and an aluminum dark slide completed the film cartridge holder. Black paint on all interior surfaces reduced flare. Exposures can be made in a darkened room simply by withdrawing the dark slide and using an opaque card to shutter the light beam.*

Cameras with Integral Lenses

The simplest way to make a photomicrograph is to use a conventional camera over the microscope. It might be an inexpensive fixed-focus camera or an expensive 35 mm camera, but it is usually one designed for regular photography of people and places. The lens is an integral part of the camera and often cannot easily be removed. A fixed-focus camera is the simplest, having only one shutter speed and usually only one lens aperture. The more expensive type of camera offers a range of shutter speeds, various distance settings, and a variety of aperture settings. This type, obviously, offers more versatility, particularly in exposure control.

When you focus a microscope visually with a normal, relaxed eye, the image may be considered to be at infinity. (Actual studies made with beam splitters used to determine the focal length of the microscopist eye while the microscope is in use have shown that accommodation varies the focal length over a wide range, and thus the actual image focus.) Therefore, the distance setting on the camera should be set at infinity. If the camera is placed over the microscope in the correct position, the image will be in focus on the film plane. The correct position of the camera lens is with the front surface of the lens at the eyepoint. Many persons can relax their eyes by staring at a very distant object just prior to focusing the microscope. Spectacles for aiding distant vision or astigmatism should be left on. If photographs are unsharp, you may be focusing your eye at some other virtual distance such as 3 or 4 metres (or 15 or 20 feet). To determine if this is the case, make a series of photographs that run the gamut of focus settings on the camera. The sharpest record will indicate the virtual distance at which you are most likely to focus the microscope. Those who cannot focus the microscope consistently will have to adopt a camera with a ground glass and, preferably, with a detachable lens.

The lens-aperture settings (*f*-numbers) on the camera *do not control exposure* as they do in regular photography. They have *no* effect on image brightness. The largest aperture setting should be used. The effect of using smaller apertures will be to vignette the image, that is, to reduce the illumination at the edge of the field because of the restrictive action of the aperture. A stopped-down diaphragm cuts into the image field and reduces field size, so only a small, circular image is recorded on the film. Actually, with

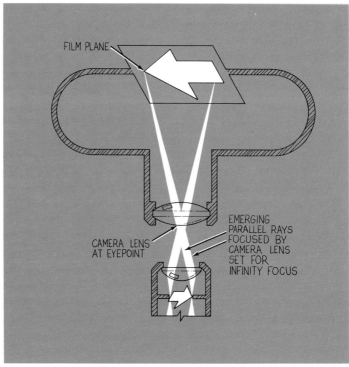

Fig. 4–3
CAMERA POSITION *for a camera with integral lens is with the camera lens at the eyepoint of the eyepiece.*

most camera lenses, even at the widest aperture there will be some vignetting. Long-focal-length lenses will eliminate the problem, but one may have to go as high as 200 mm before the vignetting is completely eliminated. On the other hand, many people consider the circular image more realistic for the microscope and do not find it objectionable.

The camera should be positioned over the microscope so that the eyepoint of the eyepiece is at, or very near, the front surface of the camera lens, as shown in Figure 4-3. This minimizes vignetting since it places the entrance pupil of the camera at the exit pupil of the eyepiece. The position of the eyepoint can be determined by holding a piece of white paper right on top of the eyepiece, then slowly raising it. A bright circle will appear on the paper. This circle becomes smaller and then larger. The position at which the circle is the smallest is the eyepoint. The distance of the eyepoint above the eyepiece will vary with different eyepieces. If the eyepiece is changed, therefore, the eyepoint may appear in a different position. The distance may only be a few millimeters or it may be almost 20

millimetres.

The camera can be held in place over the microscope by any available means. You can construct a vertical stand out of wood or metal, use a laboratory ring stand, attach the camera to the upright member of a small enlarger (with the enlarger head removed), or use a camera tripod. An elevator tripod with the elevator portion inverted is useful. In any case, the support should hold the camera firmly in the correct position. You should still be able to move the camera up out of the way or swing it to one side to look into the microscope and adjust image focus. The rule is: Focus the microscope; do not change the distance setting on the camera.

A piece of black cloth or tape or front-to-front mating (different size) sun shades can be used to exclude light between the eyepiece and camera lens. Place one shade on the camera and the other on the microscope.

When the image appears sharp, the camera can be replaced in position and the camera shutter actuated to make an exposure. An arrangement must be made to bring the camera back to the correct position. Some kind of positioning stop on the stand should be devised.

Exposure becomes a problem with simple, fixed-focus cameras because usually only one or two shutter speeds are available. This means that you must have a very bright image in the microscope and, of course, a very bright light source. The exposure time, or shutter speed, is usually very short and will be about 1/30 or 1/40 second.

Magnification With a Simple Camera

By convention, the magnifying power of optical components is based upon the normal close-focusing distance of the eye—250 mm (10 inches). When an image is formed outside the microscope, the magnification, as seen in the microscope (objective power times eyepiece power), is reproduced only if the image is 250 mm (10 inches) from the eyepoint. An integral camera lens always has a focal length *shorter* than 250 mm. Image (lens-to-film) distance becomes the determining factor for magnification of the photographed image. A camera with a 50 mm lens will record an image only about one-fifth the size of the image seen in the microscope. Magnification on film can then be determined by the ratio between the camera-lens focal length and 250 mm, multiplied by the visual magnification of the microscope. Camera lenses, except those on inexpensive cameras, usually have their focal lengths engraved on their mountings. Fixed-focus cameras normally have no designation for either focal length or lens aperture.

To determine the actual magnification recorded on film, you can place a stage micrometer slide on the microscope stage. Focus sharply on the lines of the slide and record the image. When the film is processed, you can measure the separation of the recorded lines and compare the measurement with the actual separation of the lines of the micrometer slide. For example, if two lines on the slide were 0.01 mm apart and the same recorded lines were 0.5 mm apart, the recorded magnification would be 50 x (0.5 divided by 0.01).

There are some disadvantages to using a camera with an integral lens in photomicrography. One is that the entire microscope field may be smaller than the film frame. The out-of-focus, peripheral area of the microscope field is then recorded. This condition can be alleviated to some extent by using high-magnification eyepieces so that only the central, best-corrected part of the microscope field will be photographed.

Complex camera lenses will sometimes create internal reflections due to multiple-element construction. These reflections reduce image contrast in photomicrography. Because simple, inexpensive cameras have fewer lens surfaces to reflect light, they will often produce better photomicrographs than expensive, complex cameras. Of course, a simple camera must be used correctly.

Some camera firms make microscope adapters that can be used to place a camera with an integral lens in correct position over a microscope. Such adapters are often made, however, for adjustable-focus cameras, not for the fixed-focus type. Camera manufacturers should be consulted for availability of this accessory for their cameras.

Cameras Without Integral Lenses

Reflex Cameras

A single-lens-reflex (SLR) camera can be adapted for use over a microscope. Many firms that manufacture reflex cameras also offer microscope adapters. Normally, when a reflex camera is to be used over a microscope, the lens is removed from the camera and one or more extension tubes are placed on the camera in the lens position. A microscope adapter ring, containing the microscope eyepiece, is then fastened in the front extension tube. The whole assembly of camera, tubes, and adapter can then be placed on the microscope, fitting the microscope eyepiece and the adapter ring directly into the draw-tube of the microscope. This assembly, in some designs, can be attached to a rigid stand that is supported independently of the microscope.

The micro-image is usually focused by adjusting the focus knob on the microscope while viewing the image in the camera's viewfinder. A better way is by adjusting the eye lens of a photographic eyepiece or by changing the mechanical tube length. With these methods, you can avoid changing the objective focus position with the focus knob after the microscope has been focused visually. Focusing on a ground glass within the viewfinder is all right for lower magnification, but beyond about 80X critical focus cannot be obtained on the ground screen. Critical focus of fine detail is difficult to achieve on a ground glass because of the coarseness of the ground surface. If a clear area is present on the ground glass near the center, this disadvantage can be overcome; but a simple clear area in a ground glass screen can lead to out-of-focus photomicrography

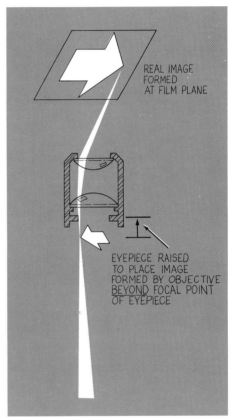

Fig. 4-4
RAISING THE EYEPIECE
allows focus of the image at the film plane of an attachment camera without lens.

because there is no reference point. Best is a ground glass with a clear center spot and *cross hairs*. Cross hairs will prevent focusing the aerial image in a plane other than the glass and allow parallax focusing. For parallax focusing, there should be no apparent motion of focus structure with respect to the cross hairs when the eye is moved slightly from side to side. Cameras with interchangeable viewing screens are most easily adaptable for photomicrography.

Bellows Extension Camera

Some firms that manufacture reflex cameras also offer an adjustable bellows, which is normally used in close-up photography. This bellows (with no lens attached) can be used on a reflex camera over a microscope and has the advantage of adjustability, so that magnification and the amount of recorded field can be varied for control of image composition. When the camera is used in this manner, it is advisable to attach the bottom plate on the bellows rack to a rigid vertical stand.

With a reflex camera, magnification can be continuously varied, a wide range of shutter speeds is available (usually 1 second to 1/1000 second) for exposure control, and the entire film frame in the camera can be filled. Also many modern reflex cameras include a behind-the-lens metering system for monitoring the image brightness in exposure determination. There are only two drawbacks, or factors, that can affect recorded image quality. One, already mentioned, is the difficulty of critically focusing the image on a ground glass. The other is the possibility of creating vibration when a focal-plane shutter is actuated. This can be minimized by using a shock mounting pad under the microscope. (See also **Camera Vibration,** page 46.) A card in the light beam can be used as an alternate shutter to minimize vibration. With the card in the light beam, (1) open the camera shutter, (2) remove the card, (3) time the exposure, (4) replace the card, and (5) close the camera shutter. Long exposure times (several seconds or more) also reduce the effect of vibration, but they do not eliminate vibration itself.

Photomicrographic Cameras

There are several commercially made cameras designed specifically for photomicrography. The most popular are those called *eyepiece cameras* or *photomicrographic attachment cameras*, or *micro attachment cameras*. A feature generally common to cameras in this

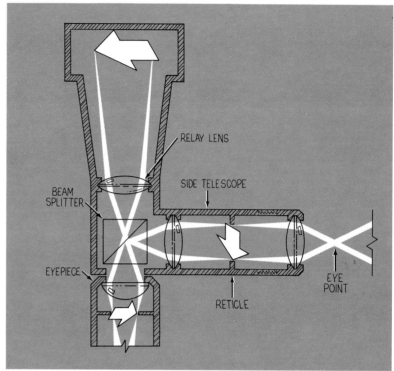

Fig. 4–5
ATTACHMENT CAMERAS
usually have a relay lens for focus at the film plane, a beam splitter, and a side telescope which is parfocal with the film plane.

group is a beam-splitter eyepiece. The image formed in the microscope can be viewed, focused, and composed by means of this auxiliary telescopic eyepiece. A central, rectangular area is often shown in the center of the field to indicate the portion of the field that will be recorded on film (Figure 4-5). Sometimes several fields of view are indicated by the reticle or graticule when the attachable photomicrographic unit is supplied with interchangeable tubes of different lengths containing different auxiliary projection lenses. In one case the largest format of three seen indicates the microscope magnification multiplied by 0.32 (a 1/3 magnification or a 3X reduction). The middle format indicates 0.5X and the smallest indicated format represents a 1X position (total microscope magnification, without reduction). The 1/3X adapter records on film most of the eyepiece field of view, the 1/2X adapter records the center portion, and the 1X adapter records only a small center section.

35 mm Eyepiece Cameras

Cameras of this type usually accommodate 35 mm film magazines, although some use sheet films (or plates). Some with more versatility are designed for use with either roll or sheet film.

The film plane in a 35 mm eyepiece camera is in a fixed position, usually far enough from the microscope eyepiece so that only the central area of the field is recorded. This feature avoids the out-of-focus peripheral area caused by curvature of field, which is inherent in many microscope objectives and which would be recorded in a camera with integral lens. The distance from eyepoint to film may vary slightly for 35 mm eyepiece cameras of different manufacture, but is usually about 100 to 125 mm. Since the distance is less than 250 mm, however, a simple correcting lens is incorporated so that the image seen sharply focused in the viewing eyepiece will also be in sharp focus in the film plane of the camera. The use of this correcting lens also allows the microscope to be used at the correct optical-tube length.

Focusing this eyepiece camera requires care. The reticle (or cross hairs) in the telescopic eyepiece must be in sharp focus when the specimen is in focus. Younger persons (usually under 40 years) may accommodate a focus difference without realizing it. The best method is to focus the reticle or graticule *first*. Do this with the focusable eye lens of the side telescope. The eye lens of the side telescope is adjustable and graduated. Look through the side telescope with the eye relaxed. Try to look *through* the graticule rather than *at* it. Start with the eye lens in the fully out position, turn the lens until the reticle is in focus, and then leave the lens in that position. If more than one person uses the same microattachment camera, record your setting of the focusing eye lens so that you can reset it quickly. Now focus the specimen either with the microscope fine adjustment or by moving the microscope eyepiece (changing the mechanical tube length) until both graticule and specimen are seen in sharp focus.

Focusing of the microscope for photomicrography is actually more critical with low power objectives. On page 16, the table shows variation of image depth with objective magnification. As the table demonstrates, slight errors in focusing with low power objectives can lead to out of focus photomicrographs. The problem arises with accommodation of the eye and failure to get the reticle and specimen in the same plane of focus. Slight errors in focus are effectively masked at higher magnifications by the greater depth of focus available.

A method for overcoming such focusing error has been suggested by several authors. The method simply involves use of a low power (about 3X) auxiliary telescope to aid in focusing. To use the telescope, first, focus it on a distant object. Place the focused telescope at the adjustable eyepiece of the microscope and focus the eye lens until the reticle is in focus. Then focus the specimen using the coarse focus of the microscope. (Do this without hesitation since accommodation of the eye may occur if focusing is done slowly with the fine focus.)

Figure 4-6 shows a typical micro-attachment camera with side telescope. The slit in the lowest attachment clamp allows focusing of eyepiece graticules when it is desired to record a scale, for example, superimposed on the specimen. On this attachment, shutter speeds are set manually, based on light-meter readings taken through the side telescope.

Readings can also be taken with the camera's own built-in metering system. Frequently, adapters are made available so that a variety of different camera models can be used with the same photomicrographic body.

Eyepiece cameras are convenient and economical, especially to provide 35 mm slides for projection purposes. This type of camera can be used with most microscopes. It requires little storage space when not in use and can be set up quickly when needed. Any efficient visual microscope setup can be used for photomicrography with an eyepiece camera. This great convenience has made the eyepiece camera very popular and useful for photomicrography.

Roll-film cameras, including 35 mm cameras, do have some limitations and disadvantages. For example, 35 mm film provides only a small-size negative. For prints, enlargement is necessary. In enlargement, the optical condition of *empty* magnification can occur. See page 14.

The beam splitter must also be considered. In some photomicrographic units the beam splitter can be swung out of the way just before the shutter is released. This increases the amount of light transmitted to the film by removing an unnecessary optical component (about 25 to 30 percent of the light goes to the side telescope when the prism is left in). On the other hand, if moving objects such as microorganisms must be recorded at a particular instant with electronic flash, or continuously on motion-picture film, it is essential to have a beam splitter that remains in place. The most versatile photomicrographic cameras offer both options for use at the photomicrographer's discretion.

Today, these simple, and generally very reliable, photomicrographic attachment cameras are offered with built-in automatic exposure devices and frequently with an automatic film advance as well. Figure 4-7 shows a typical automatic-exposure photomicrographic camera and motorized camera back, together with the necessary electronic controls.

Notice the dark slide in the film transport back in Figure 4-7. Naturally this dark slide must be pulled out to the open position before a photomicrographic is made. Be careful! It is a common error to forget to pull the dark slide out. One photomicrographer suggests hanging the end flap of the film carton on the dark slide handle via a Christmas tree ornament hook whenever the dark slide is closed. The bright film-carton end then serves to remind the photomicrographer to pull the dark slide out before resuming photography. (It is also a reminder of the type of film in the

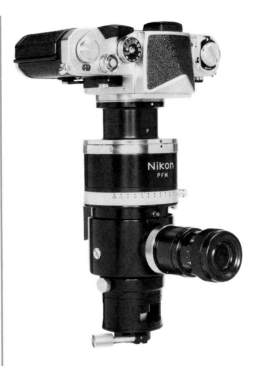

Fig. 4-6
PHOTOMICROGRAPHIC ATTACHMENT CAMERA.
Photo courtesy Nikon, Inc.

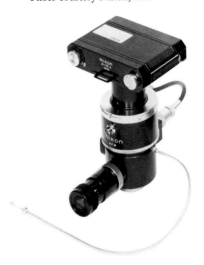

Fig. 4-7
AUTOMATIC ATTACH-MENT CAMERA *and controls.*
Photo courtesy of Nikon, Inc.

camera.) The advantage of the dark slide is that the back alone can be removed without exposing film. This allows removal of a short length or exposed film in the darkroom or replacement of one back with another back containing a different film.

Sheet-Film Eyepiece Cameras

Sheet-film cameras of the eyepiece type often reproduce visual microscope magnification, since the eyepiece-to-film distance can be fixed at 250 mm (10 inches). In addition to an observing eyepiece, these cameras often have a ground-glass back to facilitate both composing and focusing of the image to be recorded. Film size is usually 4 by 5 inches. The same field can be easily exposed on several different emulsions and then individual sheets can be developed appropriately. Adapter backs are available to convert sheet film cameras to the use of roll film or for instant print materials. These larger format cameras spread the same light used for 35 mm photography over a much larger image area; therefore, exposure times must be longer.

Fig. 4-8
TRINOCULAR ATTACH-MENT CAMERA. *This arrangement allows flexibility in substitution of components.*

Trinocular Attachment Cameras

Another type of photomicrographic camera is shown in Figure 4-8. This type is especially designed by some manufacturers for use on their trinocular microscopes. A trinocular microscope has a binocular arrangement for visual work and a third tube for photomicrography. The visual and the photomicrographic optical systems are parfocal; that is, the image seen visually in the microscope will also be in focus in the camera. A 35 mm back and a sheet-film back are interchangeable. Also, a special *field of view* eyepiece is supplied, containing a focusable eye lens and a graticule to indicate the area of field that will be recorded.

Although this is an excellent system for photomicrography, it may have one disadvantage, depending on the design. It may not allow simultaneous viewing and photography. A prism is used to change the light path from the visual system to the photographic, so it is impossible to see the field while a photograph is being made. If only stationary subjects are to be photographed, there is no problem. The eyepiece camera, on the other hand, can be used to photograph *both* stationary and moving subjects while viewing the field.

In other designs, however, there are several prism positions, including a beam splitter arrangement just as in the eyepiece camera. This permits both stationary and moving subjects to be photographed. These trinocular arrangements allow for attachment of both large format cameras and manual or automatic 35 mm cameras. Examples of automatic 35 mm cameras without side telescopes mounted on trinocular microscope bodies are illustrated in Figure 4-9.

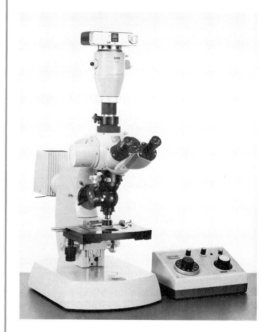

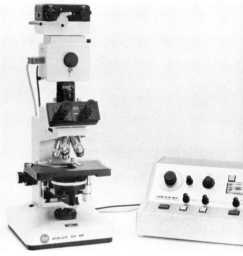

Fig. 4-9
TRINOCULAR MOUNTING *of 35 mm cameras without side telescope.* (a) Photo courtesy of Carl Zeiss, Inc. (b) Photo courtesy of E. Leitz, Inc.

Cameras With Adjustable Bellows

A camera that has an adjustable bellows and accepts a large film size is considered best for high-quality photomicrography by many microscopists. Such a camera usually contains a shutter capable of a wide range of exposure times, a light-locking device to exclude all light from the film except that from the microscope, and a ground-glass screen for focusing and composition. *No* lens is used or needed in this camera, although a special photographic eyepiece for the microscope is recommended. This type of eyepiece will have an adjustable eye lens that may be graduated with the projection distance. In use, the distance between the eyepoint (Ramsden disc) and the film plane is measured, and the measurement is set on the adjustable eye lens opposite the reference mark. Sheet films or plates in appropriate holders are normally used, but the camera will often accommodate roll film or instant-print adapter backs. Film size is commonly 4 x 5 inches or larger to allow highest magnifications and greatest field sizes.

A camera with adjustable bellows is capable of a complete and continuous range of magnification, since eyepiece-to-film distance is adjustable. Remember that microscope visual magnification is reproduced at a 250 mm (10-inch) distance.

Commercially made bellows cameras are usually mounted vertically on a rigid stand (Figure 4-10). A heavy, rigid stand is necessary in order to avoid vibration and consequent unsharpness in the image. In addition, use of a separate means of absorbing vibration and shock, such as a vibration isolation table, provides further stability, particularly when long exposure times are indicated.

A precaution that should be observed with a separately-mounted bellows camera is correct positioning of the shutter. Ideally, the shutter should have a reasonably large opening and small, light, fast-moving blades. It should be positioned so that the blades are at, or very near, the eyepoint of the microscope eyepiece. If the blades are too far above the eyepoint and short exposure times are used, a faint silhouette-image of the shutter blades, called *shutter shadow,* may be recorded.

One of the versatile features of many 4 x 5 bellows-camera systems is that the lower part of the bellows can be raised away from the trinocular tube so that a 35 mm eyepiece camera can be quickly attached when 35 mm slides are required.

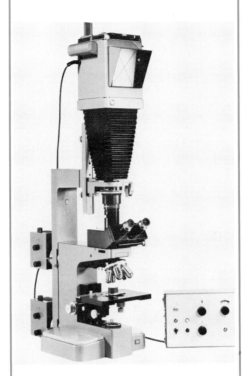

Fig. 4–10
RIGID STAND *is part of this commercially integrated bellows camera.* Photo courtesy E. Leitz, Inc.

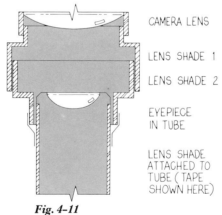

CAMERA LENS

LENS SHADE 1

LENS SHADE 2

EYEPIECE IN TUBE

LENS SHADE ATTACHED TO TUBE (TAPE SHOWN HERE)

Fig. 4–11
NESTING LENS HOODS *serve as light lock between a separate camera adapted for photomicrography and the microscope eyepiece.*

Flare, Focus, and Vibration

Light Lock

Commercial photomicrographic cameras of the adjustable-bellows type usually contain an efficient system for excluding extraneous light from the camera. This is the light lock—the connection between microscope and camera. When an available view camera or an enlarger is adapted for this work, a light-lock must be devised. A very efficient and convenient light lock can be made by using two different-size lens hoods—the smaller one is attached to the camera or shutter, and the larger, to the tube of the microscope. The two hoods will nest together to exclude all light from the camera except that coming through the microscope. The hoods and any adapter rings should be thoroughly blackened.

In practice, the camera is usually moved out of position so that the microscope can be used visually to locate an appropriate field. The camera is then moved into position so that its lens hood fits into the lens hood on the microscope. The camera, of course, should be centered with respect to the microscope so that there is no physical contact between the two lens hoods. Enough space exists to allow movement of the microscope barrel for critical focusing of the image on the ground glass.

The Ground Glass

Frequently in photomicrography the image to be recorded consists of extremely fine detail. The surface structure of a ground-glass screen, however, is often so coarse that it interferes with this detail; critical focus of the image is difficult, if not impossible. Some manufacturers of photomicrographic equipment provide screens with clear centers for critical focus of the image on the plane of the ground surface. A ground-glass screen with clear center can also be prepared in the following manner.

With a soft pencil, or better, India ink pen, draw a cross or two diagonals on the ground-glass surface that will be facing the microscope. Place a small drop of warm Canada balsam (or other mounting medium) where these diagonals cross. Drop a small, round, micro cover glass onto the balsam and press it into place gently. Allow the balsam to set. A permanent transparent center on the ground glass will result. This procedure provides cross hairs for parallax focusing.

For photomicrography, place a mounted hand magnifier of at least 5X over the clear center of the screen. Put your eye to the magnifier and adjust it for sharp focus of the ruled cross lines. Then focus the specimen and glance slowly back and forth over the crossed lines. When the image is critically focused in the plane of the lines, there will be no movement of the image in relation to the crossed lines.

Camera Vibration

Vibration in a camera setup can result from microscope manipulation, from drawing and replacing dark slides in film and plateholders, from setting the mechanical shutter, and from transmission through the stand. It is always good practice to wait a few seconds for vibration to cease before making the actual exposure. Even then, if the shutter is actuated manually, additional vibration can occur and may cause some unsharpness in the recorded image. If the shutter is part of the camera, use a suitable cable release to avoid vibration from this source. A self-timer also provides a means of automatic delay before the shutter is operated.

Some cameras contain focal-plane shutters; others have leaf-type, or *between-the-lens,* shutters. It is generally believed that the latter type is much more satisfactory for photomicrography since a focal-plane shutter may impart a certain amount of vibration when an exposure is made. Actually, this belief assumes shutters of comparable make and quality. Vibration is caused by the abrupt start and stop of the focal-plane shutter curtain as the shutter opens and closes. This shock can be alleviated by using longer exposure times to minimize the vibration effect caused by the shutter or by using special absorbing pads or an antivibration table under the microscope. These will dampen the vibrations; tests will determine their efficacy. Such units are available from scientific equipment supply firms.

There is at least one setup in which the photomicrographer has the choice of leaf or

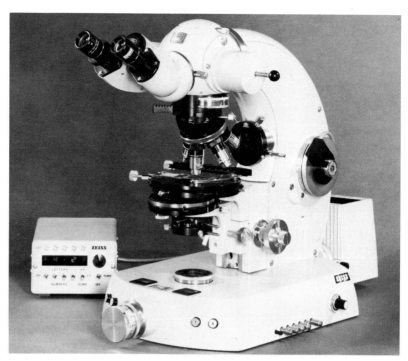

Fig. 4–12
PHOTOMICROSCOPES *include integrated camera and automatic exposure calibration. The complex internal light path of the microscope pictures (a) is shown in the drawing (b)* Photos courtesy of Carl Zeiss, Inc.

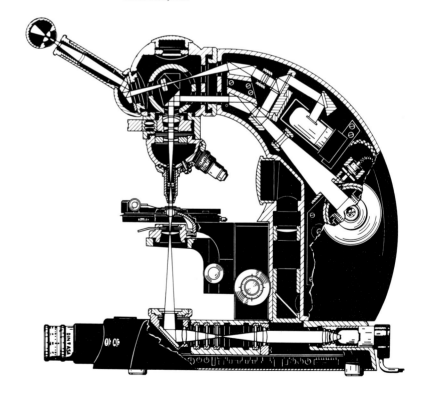

focal-plane shutter. In this particular case, the focal-plane shutter is of much higher quality than the leaf shutter, and photomicrographs of test specimens made with it are sharper than those made with the leaf shutter. Actual tests should always be made when a choice of shutter type and location is available.

In general, heavy shutters and focal-plane shutters on the camera should not be used for timing the exposures. It is better to set them to the "open" position just prior to making the exposure and then to time the exposure with an auxiliary shutter placed in the light beam between the lamp and the condenser. But be careful to keep the shutter as near as possible to the microscope condenser because if it is near to the field diaphragm, it will be imaged in the field of view. Control of light with a supplementary shutter in the light path and the camera shutter open is awkward, but it does result in vibration-free photomicrographs.

Photomicroscopes

There are commercially available instruments that are made specifically for photomicrography or for microscopical work that requires frequent photomicrographic documentation. These instruments are photomicroscopes or photographic microscopes in which the photomicrographic apparatus is built in rather than attached as an accessory. Figure 4-12a shows the Zeiss photomicroscope III with automatic data-recording unit. The 35 mm film is held in the cylinder device with the round cover at the rear of the body, and all photographic controls are built in. Figure 4-12b shows the light path through the photomicroscope. After the light passes through the specimen and objective, it encounters a beam splitter where most of the light passes through to the film. The remaining, deflected light encounters another beam splitter that directs a portion of the light to the built-in photomultiplier tube, and the remainder to the eyepieces.

The inverted camera microscope shown in Figure 4-13 contains both fully automatic 4 x 5-inch and 35 mm camera systems.

Choice of Camera

We have seen that photomicrographs can be made with no camera at all, with simple cameras that have fixed lenses, with more complex cameras that have removable lenses, with attachment 35 mm and large-format eyepiece and bellows cameras, and with microscopes having built-in photomicrographic capabilities. Costs range from that of film alone to many thousands of dollars. The choice of a specific method and type of equipment must be made by each individual photomicrographer, taking into consideration such things as anticipated photo work load, variety of microscopical tasks, cost, personal preference, and experience. The principles for producing the best visual image are the same for all microscopes; the means of recording the image is optional. The simple camera, attachable photomicrographic unit, and built-in photo system all are capable of producing high quality photomicrographs.

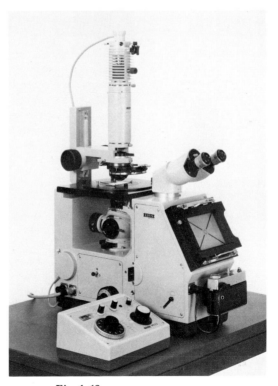

Fig. 4-13
INVERTED CAMERA MICROSCOPE *provides for 35 mm or 4 x 5 inch camera modes.* Photo courtesy of Carl Zeiss, Inc.

Chapter Five
COLOR PHOTOMICROGRAPHY

The ultimate function of the microscope and its illumination is to produce the best possible image of the specimen. If the specimen is colored, as with biological stains, then the image will also be colored; and it can be recorded to best advantage on a color film. Filters are almost always necessary when a color film is to be exposed through the microscope, principally for control of color balance.

A working knowledge of color film characteristics, as well as of the action of specific filters, is essential in color photomicrography if the microscope image is to be recorded effectively.

Color Films

There are two general types of color films for exposure in a camera. They are *reversal* color films and *negative* color films. The reversal films yield direct-positive color transparencies after reversal processing; the resultant colors are comparable to those seen in the original subject. Color transparencies are viewed either by projection or on a suitable illuminator. The negative films yield color negatives after processing; the colors in the recorded image are *complementary* to the corresponding colors in the subject. Color negatives must be printed onto a color print material to obtain a reproduction of the original subject colors.

Color films are further identified by the kind of illumination for which they are balanced in manufacture. This is either artificial light produced by tungsten lamps, or daylight. The designations Type A, tungsten, or Type B, indicate that the tungsten light source for which the film was balanced had a color temperature of 3400 K (Type A) or 3200 K (tungsten and Type B). A film specified as daylight type is balanced for average sunlight or electronic flash at 5500 K. Kodak professional films are designated either Type L or Type S. Type L (for long exposure) film is balanced for tungsten light at 3200 K and for long exposure times (1/10 second to 60 seconds). Type S (for short exposure) film is balanced for daylight illumination and for short exposure times of 1/10 second or less.

When a particular film is to be exposed in photomicrography and the illumination differs from that for which the film was balanced, specific filters must be placed in the light beam to adjust the illumination. KODAK Light Balancing Filters were made for this purpose. Failure to make this adjustment may result in incorrect color balance in the recorded image.

All color films are rated according to their speeds, or sensitivity to light. Film speeds are assigned by a standard rating system as numbers that are directly proportional to sensitivity—the larger the number, the higher the sensitivity. A film with an arithmetic speed value of 160 is twice as fast in its reaction to light as another film rated at 80, four times as fast as a film rated at 40. The speed number commonly given is the ASA speed or, more recently, ISO speed.

Selection of Color Films

The primary characteristics to consider in selecting a color film for photomicrography are film format, film speed, and the ability of the film to record specimen colors as accurately as possible. Other factors are contrast, resolving power, granularity, illumination, and color balance. *(The appendixes contain more information on film selection.)*

The film size, of course, is governed by the camera to be used. Most photomicrographic cameras accommodate 35 mm film to produce a 24 x 36 mm frame size. Others accommodate sheet films of specific size. A wide selection of color films is available in 35 mm and sheet-film sizes.

In photomicrography, 35 mm reversal color films are most often used. These films yield color slides commonly used in 2 x 2-inch slide projectors. When these films are processed, the individual frames are usually mounted either in cardboard or plastic or bound in glass mounts for direct insertion in the projector. For handling, filing, and transporting, the convenience of 35 mm slides is unquestioned.

The use of a high-speed film may be of advantage when a specimen is in motion or possibly, when the illumination intensity is very low as in fluorescence photomicrography. However, microscope subjects normally are stationary, and film speed is not of primary importance because exposure time can be controlled over a wide range.

Either daylight or tungsten-balanced films can be used successfully in photomicrography if the illumination is properly adjusted with light-balancing filters. Normally, however, if tungsten illumination is used with the microscope, a color film balanced for tungsten should be used. Daylight-type films are used when the light source has daylight quality; such sources include xenon arc or electronic flash. Color films balanced for daylight can also be used with tungsten sources if the illumination is correctly adjusted. Introduction of conversion filters reduces light intensity and therefore increases required exposure time. The use of daylight films extends the selection of films available when one is looking for a film that will record and reproduce specimen colors most accurately. The rendition of specimen color is related to the sensitivity and spectral response of the color film. (See Figure 5-1, page 56.)

Instant color materials provide immediate results when working at the microscope. Microscope manufacturers provide for use of instant print materials with adapters for their microscopes. KODAK Instant Color Film can be used with any microscope equipped with the appropriate manufacturer-supplied adapter or with a KODAK Instant Film Back fitted via a 4 x 5-inch Graflok® back. The KODAK Instant Film Back is motorized and offers an optional 110-volt power source. KODAK Instant Color Film is daylight balanced and has sufficient speed to allow for any required filtration of the light source. (See Fig. 5-2 and Fig. 5-3, page 56.)

Rendition of Stain Color

The many biological stains used in specimen preparation, such as tissue sections and smears, represent all of the colors that can be seen. Different color films show some differences in sensitivity and response to various colors. For this reason, the color of one stain may record better on one color film than another. It depends on what that color is and how sensitive the particular color film is to that precise color. The rendition of a stain color also depends on the dyes formed in the emulsion layers of the film; different color films utilize different dye systems. So, in photomicrography, specimen dye colors are recorded on a color film containing different dyes. Unfortunately, the results are not always as expected.

Although a comprehensive survey has not

been made on the rendition of all biological-stain colors on the different color films, there are some common stains whose color rendition is known. Probably the most common stain is *eosin.* This reddish stain is widely used with tissue sections in animal histology in combination with *hematoxylin,* and in combination with *methylene blue* stain for blood smears *(Wright's stain).* The use of a *didymium* glass filter enhances the color rendition of these stains with certain films.* Newer EKTACHROME Films (Process E-6) display enhanced response with the didymium filter and some histological stains. Filtering can often yield an improvement in the representation of many other stains.

Processing Color Films

After color films have been exposed in a camera, they must be properly processed, either by the user or by a commercial processing laboratory. Many commercial laboratories are equipped to process and print Kodak color films. Some of these processing laboratories offer specialized services, such as push-processing of films for higher film speed and contrast.

Some Kodak color films can be processed by the user. They include KODAK Photomicrography Color Film 2483 (Process E-4),† KODAK EKTACHROME Films (Process E-6), KODACOLOR Films (Process C-41), and KODAK VERICOLOR Professional Films (Process C-41). Chemicals are available in kit form for each color process. Complete information is given in data sheets about mixing chemicals, processing times, agitation, and temperature.

KODACHROME Films cannot be processed by the user. They must be processed by a facility that has the elaborate equipment required for this purpose.

Increasing Color-Film Speed

It is possible to use KODAK EKTACHROME Films (Process E-6) at film-speed ratings higher than normal by increasing the first-developer time a specific amount. These films can be processed with Process E-6 chemicals. Increasing film speed by altering processing is not recommended for other Kodak color films.

Best color quality will be obtained when color films are processed normally. However, with many photomicrographic subjects, increasing film speed of KODAK EKTACHROME 50 Professional Film

*Kodak does not supply didymium filters. They are available from microscope suppliers. A typical filter is the Schott didymium multiband glass filter BG20 in the 1 or 1.5-mm thickness.

(Tungsten) one stop and increasing processing time gives no apparent diminution of quality, and the resulting contrast increase is valuable in the rendition of many subjects (see Fig. 5-4, page 56).

Processing for increased speed can easily be accomplished when users develop their own color films. If film is usually sent to a commercial laboratory, users should check with the laboratory to find out if this service is available.

Filters

The ultimate function of both the microscope and the illumination is to produce the best possible image of a specimen. A photomicrograph is a record of this image on a photographic material. If the specimen is colored, as with biological stains, then the image will also be colored and will be recorded most successfully on a color film. If a colored image is to be recorded on a black-and-white material, the colors must be reproduced as tones of gray that satisfactorily represent the color brightnesses in the specimen. It is usually necessary to use specific light filters either to provide correct rendition of colors on color film or to record the colors as appropriate gray tones with suitable contrast on a regular film or plate.

Light Balancing Filters

These filters are intended for the adjustment of illumination when it differs from that for which a color film is balanced. Illumination color quality is usually expressed as *color temperature* in Kelvin (K). KODACHROME 40 Film 5070 (Type A) is balanced in manufacture for a light source having a color temperature of 3400 K. If the light source to be used in exposure of this film has a color temperature *lower* than 3400 K, its illumination can be adjusted with one or more KODAK Light Balancing Filters of the 82 series to approximate a 3400 K color temperature. There are four filters in this series: No. 82, 82A, 82B, and 82C. Each will effectively raise color temperature by definite increments. Of course, these filters don't *actually* change the color temperature of the light source, but they do modify the illumination to simulate a higher color temperature. The No. 82 filter will effectively increase the color temperature by 100 K; the No. 82A, by 200 K; the No. 82B, by 300 K; and the No. 82C, by 400 K. If the light source had a color

†Availability of processing services for Process E-4 is now limited. Processing services or chemicals for user processing of Photomicrography Color Film are still available from Kodak.

temperature of 3000 K, therefore, its illumination could be adjusted to 3400 K with a No. 82C filter. The above filters can be used in combination to adjust illumination by greater amounts than 400 K. For example, if the light source were 2800 K, its illumination could be adjusted to 3400 K with a No. 82C plus a No. 82A filter. All of the 82 series filters are light blue in color. Since virtually every microscope illuminator that is not xenon or electronic flash has a color temperature less than 3400 K—usually 2700 K-3200 K—filters of the 82 series are almost always used in the light path of the microscope.

Another group of KODAK Light Balancing Filters applies when color temperature is too high for a film. This is the No. 81 series, which includes No. 81, 81A, 81B, 81C, 81D and 81EF. These filters are yellowish in color and will effectively *decrease* color temperature by 100 K, 200 K, 300 K, 400 K, 500 K, and 600 K. These figures are only approximate since the actual amount depends upon the initial color temperature. The No. 81 series filters are not often used in photomicrography since too high a color temperature seldom occurs. However, a common use of the 81 series is to slightly lower the color temperature of electronic flash.

An important fact to be considered is that the actual color temperature of light source is usually unknown. Also, the illumination quality at the film plane may be different from that emitted from the source because of optical absorptions within the microscope and the illumination system. Most built-in illumination systems, for example, contain a diffuse surface, which will effectively *decrease* color temperature by about 200 K. So, even if the source were known to have a specific color temperature, the illumination color-quality supplied to the film might be different. Most often, it is *lower* in effective color temperature than is indicated by the source.

Tungsten-filament lamps are often used in photomicrography, particularly in the built-in illumination systems where 6-volt and 12-volt coil-filament lamps are common. These lamps, manufactured by several firms, differ in color temperature through a range of about 2800 K to 3200 K, even when used at the specified voltage. A lamp in one microscope system may differ in color temperature from the lamp in another system. It is difficult to say which light balancing filter, or filters, should be used. Tungsten-halogen lamps, on the other hand, are quite consistent in their output with a nominal 3200 K color temperature.

The most practical method for determining the correct filters is to make color-balance

KODAK Filters for KODAK Color Films

Light Source	Tungsten, Type B or Type L Film (3200 K)	Type A Film (3400 K)	Daylight-Type Film
	Filter No.	Filter No.	Filter No.
6-Volt Ribbon-Filament (about 3000 K)	82A	82C	80A + 82A
6-Volt Coil-Filament (about 2900 K)	82B	82D	80A + 82B
100-Watt Coil-Filament (about 3100 K)	82	82B	80A + 82
300- to 750-Watt Coil-Filament (3200 K)	None	82A	80A
Tungsten Halogen (3200 K)	None	82A	80A
Zirconium Arc (3200 K)*	None†	82A	80A
Photoflood Lamp (3400 K)	81A	None	80B
Carbon Arc (3700 K)*	81D†	81B	80C
Xenon Arc (5500 K)*	85B	85	None*

*The suggested filters give approximate color-temperature compensation. Color test exposures are necessary to obtain best color balance.

†A No. 2B filter is often used to absorb unwanted ultraviolet from arc illumination. This is not necessary with many Kodak color reversal films intended for daylight exposure since these films have a UV-absorbing overcoat. Color films intended for tungsten exposure and black-and-white films do have UV response.

tests, using different filters or combinations of filters. The 82 series of light balancing filters usually applies with tungsten illumination since the color temperature is almost invariably too low for an artificial-light film. If the initial color temperature of the source is unknown, make a series of exposures on a reversal color film balanced for tungsten illumination using, first, no filter, then the No. 82, 82A, 82B, 82C, 82C + 82, and 82C + 82A filters (see table for exposure adjustments). After the color film is processed, examine the different exposures on a suitable illuminator to determine which one shows the most pleasing rendition of specimen colors and a clean, almost-white background. (A graphic arts illuminator of 5000 K is good for this.) The filter or filters used to make that exposure would then be correct for future exposures. If a transformer was used to provide the 6 volts to the lamp, the same setting should always apply for making future photomicrographs. Several voltage settings are often available on a transformer. The *highest* setting should normally be that recommended as normal by the bulb manufacturer. It is not necessary to overvolt a lamp and shorten its life.

If the color temperature of the source is known, even approximately, the color-balance test series can be much shorter. Very often only about two filters or combinations need to be tried.

Light balancing filters should be placed between the illuminator and the condenser of the microscope. Separate illuminators usually contain filter receptacles for the purpose, but this frequently places the filters too close to the field diaphragm. In this position, any dirt or flaws in the filter will be imaged in the plane of the specimen. It is best to put the filters closer to the condenser. Avoid putting them close to the lamp as excessive heat will damage gelatin filters.

The table above contains typical filter recommendations for different light sources and various color films. These recommendations are approximate, because of the variance in the color temperature of light sources at the time of exposure. Voltage regulators are necessary for critical work.

All daylight-type color films can be used in photomicrography, regardless of the type of illumination, if the correct filters are placed in the *light beam* to adjust the illumination to daylight quality. These films can also be used, of course, with light sources of daylight quality (such as the xenon arc or electronic flash) and probably without any light balancing filters. With tungsten lamps and the zirconium arc, however, the use of specific light balancing filters is necessary.

A blue KODAK WRATTEN Gelatin Filter No. 80A will adjust illumination color-quality from 3200 K to daylight. If color temperature is below 3200 K, as it often is, one or more bluish filters of the 82 series will also be needed. In order to establish which filter of the 82 series should be used with the

No. 80A filter, make a color test-exposure series. Five exposures will probably suffice: one with the No. 80A filter alone and four with the combinations of No. 80A + No. 82, No. 80A + No. 82A, No. 80A + 82B, and No. 80A + No. 82C. One of the resultant exposures will show the best color balance. The correct combination should be used in subsequent photomicrography when a daylight-type film is exposed.

Color Compensating Filters

Undesirable color effects are caused by several factors in color photomicrography in addition to the one already mentioned (exposing a color film to illumination other than that for which the film was designed). These color effects can usually be corrected with KODAK Color Compensating Filters, available in various densities in red, green, blue, cyan, magenta, and yellow. These filters are commonly called *CC filters*, and they are supplied as gelatin film squares.

The various general effects that can be corrected with CC filters are discussed in **Factors Affecting Color Balance,** page 52. Pale color compensating filters can sometimes be used to enhance the rendition of certain stains. However, the benefits gained by use of a filter of the approximate color of the stain must be balanced against a possible degradation of complementary stain color. A selection of CC filters is most useful for the critical photomicrographer. In practical terms, it will pay to have all six colors in densities of 0.5, 0.10, and 0.20. These can be combined to get values of 0.15, 0.25, 0.30, and 0.35 as well. (See Figure 5-5, page 57.)

While gelatin or glass filters will provide the highest optical quality and edge-to-edge uniformity, many microscopists have found that less expensive substitutes are acceptable when they are used in the illuminating beam.

One of these is the KODAK Color Print Viewing Filter Kit designed for visual evaluation of color prints. The kit has a card containing three 40 x 75 mm viewing filters with density values of 0.10, 0.20, and 0.40 for each of the six colors: red, green, blue, cyan, magenta, and yellow. These viewing filters are not manufactured to the same critical tolerances as KODAK CC or CP Filters and some unevenness of density may occur across the filter. Nevertheless, the kit filters can be useful for making experimental ring-arounds and test exposures.

Another approach is the use of acetate color printing filters in the illuminating beam. An inexpensive set of acetate filters is the KODAK Color Printing Filter Set (Acetate).

Conversion and Light Balancing Filters*

Blue Conversion Filters

Filter Number	Exposure Increase in Stops†	Conversion in Kelvin	Mired Shift Value
80A	2	3200 to 5500	−131
80B	1⅔	3400 to 5500	−112
80C	1	3800 to 5500	−81
80D	⅓	4200 to 5500	−56

Bluish Light Balancing Filters

Filter Number	Exposure Increase in Stops†	To obtain 3200 K from:	To obtain 3400 K from:	Mired Shift Value
82C + 82C	1⅓	2490 K	2610 K	−89
82C + 82B	1⅓	2570 K	2700 K	−77
82C + 82A	1	2650 K	2780 K	−65
82C + 82	1	2720 K	2870 K	−55
82C	⅔	2800 K	2950 K	−45
82B	⅔	2900 K	3060 K	−32
82A	⅓	3000 K	3180 K	−21
82	⅓	3100 K	3290 K	−10

Amber Conversion Filters

Filter Number	Exposure Increase in Stops†	Conversion in Kelvin	Mired Shift Value
85C	⅓	5500 to 3800	81
85	⅔	5500 to 3400	112
85N3	1⅔	5500 to 3400	112
85N6	2⅔	5500 to 3400	112
85N9	3⅔	5500 to 3400	112
85B	⅔	5500 to 3200	131
85BN3	1⅔	5500 to 3200	131
85BN6	2⅔	5500 to 3200	131

Yellowish Light Balancing Filters

Filter Number	Exposure Increase in Stops†	To obtain 3200 K from:	To obtain 3400 K from:	Mired Shift Value
81	⅓	3300 K	3510 K	9
81A	⅓	3400 K	3630 K	18
81B	⅓	3500 K	3740 K	27
81C	⅓	3600 K	3850 K	35
81D	⅔	3700 K	3970 K	42
81EF	⅔	3850 K	4140 K	52

*KODAK WRATTEN Gelatin Filters and KODAK Light Balancing Filters

†These values are approximate. For critical work, they should be checked by practical test, especially if more than one filter is used.

Density–Percent Transmittance Table

Density	Percent Transmittance	Density	Percent Transmittance
0.10	80.0	0.80	16.0
0.20	63.0	0.90	13.0
0.30	50.0	1.00	10.0
0.40	40.0	2.00	1.0
0.50	32.0	3.00	0.10
0.60	25.0	4.00	0.010
0.70	20.0		

Available in 75 x 75 mm squares, the filter set contains 17 filters—one each of the CP05, CP10, CP20, and CP40 values of magenta, yellow, cyan-2, and red plus a CP2B ultra-violet absorber.

Neutral Density Filters

Neutral density filters can be used in photomicrography to *reduce* image brightness as a means of controlling exposure time. A neutral filter will absorb a specific amount of light, depending on its *density*, without affecting the color quality of the illumination. The principal application of such filters is in the exposure of color films. They can also be used with black-and-white films to prevent overexposure.

If a very intense light source is used to provide illumination, the correct exposure time (as determined by a test-exposure series or by light measurement) may be shorter than the fastest available shutter speed. A neutral density filter can be placed in the light beam to reduce image brightness so that the exposure time will be within the shutter speed range.

The absorption of light by a neutral filter is directly proportional to the filter's *density*. The greater the density, the greater the amount of light absorbed. Also, the greater the density, the less the amount of light *transmitted*. * The table (below left) includes various filter densities available and their corresponding transmittances.

The neutral density filters most often used in photography to control exposure time have densities of 0.30, 0.60, and 0.90. As shown in the table, these filters have transmittances of 50, 25, and 13 percent. Since a density of 0.30 has a transmittance of 50 percent, it can be used to reduce brightness by a factor of 2. A density of 0.60 has a transmittance of 25 percent and a reduction factor of 4. These densities can be used in combination because the total density equals the sum of the individual densities.

*Density is defined as the common logarithm of the opacity. Opacity is the reciprocal of the transmittance; transmittance, in turn, is that fraction of light incident on a material that passes through the material. Thus,

$$\text{Transmittance} = \frac{\text{Transmitted light}}{\text{Incident light}}$$

$$\text{Opacity} = \frac{1}{\text{Transmittance}}$$

$$\text{Density} = \log_{10} \text{Opacity} = \log_{10} \frac{1}{\text{Transmittance}}$$

When considering transmission of light, it is appropriate to use density rather than opacity or transmittance values, because the eye judges brightness differences on a logarithmic scale.

Here is a sample exposure calculation involving neutral density filters: Suppose that a light source provides very bright illumination, and that a reasonable fast film is in the camera. The correct exposure time is 1/125 second, but the fastest available shutter speed is only 1/60 second. This shutter speed would cause overexposure. A density of 0.30 in the light beam reduces image brightness by 50 percent; 1/60 second would then be the correct exposure. If a density of 0.60 were placed in the light beam, the correct exposure setting would then be 1/30 second.

Because reversal color films have a very short exposure latitude, the best exposure might be between two shutter speeds. For example, 1/60 second may be too short, causing slight underexposure; but 1/30 second may be too long, causing slight overexposure. In this case, a neutral density filter of either 0.10 or 0.20 could be used with a shutter speed of 1/30 second to reduce the light level by 1/3 or 2/3 of a stop. (Remember that a density of 0.30 reduces light intensity by 50 percent or one full stop.) The routine use of 0.10 and 0.20 neutral filters is an excellent way of getting 1/3 stop differences in successive exposures for critical photomicrographs.

Neutral density filters are also useful for reducing visual image brightness. Place a very dense filter in the light path when image brightness is too high for comfortable viewing. A density of 1.00 or more will usually suffice for this purpose, but it should be removed for photography because it may prolong exposure time too much. Avoid long exposures when possible because of *reciprocity effects* with color films and the resultant possibility of color shift and decreased film speed.

Some neutral filters normally used to reduce brightness may have a slight yellowish color, either inherently or as a result of aging. This yellowish color can be neutralized with a pale blue color compensating filter such as a CC05B or a CC10B. The effect generally isn't very great and can often be ignored. Evaporated metal on glass is one of the best forms of neutral density filter.

Factors Affecting Color Balance

At one time or another probably everyone who has made color photomicrographs has encountered undesirable color effects. Sometimes the reason is immediately obvious or can be easily determined by reviewing the conditions that affect exposure. Occasionally,

however, the cause of erroneous color balance baffles even the most careful photomicrographer. When the cause is known, suitable correction can often be made in subsequent exposures. If it remains in doubt, a complete review of possible causes is necessary. Here are some suggestions.

Many undesirable effects can be neutralized by placing appropriate KODAK Color Compensating (CC) Filters in the microscope light beam. Some effects on color balance, however, are due either to incorrect use of optics or to problems related to film storage and film processing. Usually, these effects cannot be compensated for with filters. The only remedy is to follow recommended procedures.

The most common reasons for a color-balance shift are listed in this section. Some are peculiar to photomicrography, but many can be encountered in any type of photography in which color film is used. Only color reversal films are considered here, since color negatives may not show a color change until they are printed. When color negatives are slightly off-balance, the effects can usually be corrected in printing if they are not too pronounced.

Color-Temperature Variance

Variation in illuminant color temperature is probably the most common reason for unexpected color shifts in color photomicrography. Whenever the illumination differs from that for which a particular film is balanced, the photomicrograph will have shifted color. If the color temperature of the light source is too high, an overall cold, bluish effect will be noticed in the photomicrograph. If the color temperature is too low, the photomicrograph will be too warm and will be either yellowish or reddish yellow in appearance. The degree of color shift will vary according to the amount of color-temperature difference between the actual light source and that for which the film was balanced. An example of a large color shift is a micrograph made on daylight color film with uncorrected tungsten illumination. (See Figure 5-1, page 56.) Color-temperature variance can be corrected by placing appropriate KODAK Light Balancing Filters in the illumination beam.

These mismatches between film color temperature and illuminator color temperature usually lead to large errors in color balance. There are two other sources of color-temperature variances that are more subtle and seldom expected, and that lead to small to moderate color shifts. One source of color-temperature variance is line-voltage fluctua-

tion. It is very instructive to connect a voltmeter to the line and observe it throughout a day. Variations of 15 percent or greater are common. Most variations are random, but some are predictable, as when a large piece of machinery is turned on. In some cases, the variation correlates with daytime factory shift changes and the like. *Brownouts*, the supply of a continuous lower voltage for extended times, may occur. To eliminate these color shifts, the critical color photomicrographer should install a constant-voltage transformer in the line between the wall outlet and the microscope illuminator. Such a constant-voltage transformer can take variations up to 15 percent in input voltage and still maintain a constant output voltage. Such a transformer is a necessary investment for critical color photomicrography, but for best results the output of the transformer must match the load of the microscope illumination system.

Another, more subtle source of color-temperature variation is the microscope illuminator's own transformer. If it is of the continuously variable type it will actually be very difficult to set and reset the variable voltage adjustment knob to the identical spot each time, even if there is an integral meter. Try using a footswitch to turn the illuminator on and off, leaving the adjustment knob setting (bulb manufacturer's recommended operating voltage) alone during the color-balance test rolls.

Heat-Absorbing Filters

Many light sources—such as high-wattage tungsten lamps and xenon arcs—emit a considerable amount of infrared radiation in their illumination. The infrared is evidenced as heat and should be removed by an appropriate *heat-absorbing filter* to protect the microscope optics, the specimen, and any filters in the light beam. Some microscope illuminators have heat-absorbing glass filters built in. Unfortunately, the owner of such an illuminator may be unaware of the presence of the filter, which is usually green or blue-green. This coloration can affect the color balance of a photomicrograph. The transparency may appear too green or blue-green, an effect that can be corrected in subsequent exposures by placing a neutralizing KODAK Color Compensating Filter in the light beam. A CC filter of a complementary color is necessary. If the heat filter is greenish, a pale magenta CC filter will absorb the green and produce a neutral effect. If the filter is blue-green, a pale red CC filter is indicated.

Heat-absorbing glass filters vary in their degree of coloration; it is not possible, therefore, to assign one specific CC filter for correction. The correct filter must usually be determined by test exposures. If the heat filter can be removed from the lamp temporarily, place it on a good 5000 K transparency illuminator and view it through CC filters of complementary colors until one is found which best neutralizes the heat-filter color. Of course, permanent removal of the heat filter also solves the problem, but this may not be wise, since permanent damage to the optics, filters, and specimen may result. These problems can be overcome today by the use of interference filters of the dichroic type to reflect infrared. Unfortunately they are expensive, but they are more efficient and do not crack or break due to too much heat absorption.

Ultraviolet Radiation

Some color films are very sensitive to ultraviolet radiation, which can be recorded as blue by the blue-sensitive emulsion layer. If ultraviolet is present, as in xenon-arc illumination, it may cause a color photomicrograph to appear too blue. An ultraviolet-absorbing filter, such as the KODAK WRATTEN Filter No. 2B, should be used. The No. 2B Filter will absorb ultraviolet radiation but will transmit all visible colors. The bluish effect of ultraviolet radiation on color film is often encountered in the photomicrography of metals. Most electronic flashtubes also emit ultraviolet radiation. If a light source that emits ultraviolet is employed in photomicrography with color films, use a UV-absorbing filter such as a KODAK WRATTEN Filter No. 2B or No. 2E. Many Kodak color reversal films intended for daylight exposure have an ultraviolet-absorbing overcoat so that use of a filter is not necessary with these films.

Biological Stains

The stains used in coloring a section or smear to produce contrast between the elements of the specimen have individual properties of absorption and transmittance. Some stain colors reproduce very well when recorded on a specific color film, but others often appear quite different on film than when seen in the microscope. Eosin and fuchsin, for example, may not record well on some films. Their rendition, however, can be noticeably enhanced when a glass didymium filter is placed in the illumination beam. The thickness of the didymium filter should not be greater than 2 mm.* If it is greater, a background color may appear or other stain colors

*See footnote, page 49.

may be degraded. Remember, because of the inherent deficiencies of the dyes used in color film, it may be impossible to record accurately all colors of the specimen.

Stained Backgrounds

Occasionally when a microscope slide is prepared by staining the specimen, the background becomes colored. This background color is recorded on color film, in some cases giving an undesirable effect. The effect can be neutralized by using a pale color compensating filter in the light beam. The filter color should be complementary to the background color.

For example, this may happen with smears of bone marrow. The thick medium in which the blood cells are dispersed may absorb the eosin stain included in Wright's stain for blood smears. The background of the smear becomes slightly pink. If this color is objectionable, it can be neutralized in a photomicrograph by placing either a pale green (for magenta stains) or cyan (for reddish stains) color compensating filter in the illumination beam. The exact color density of the CC filter used will depend on the intensity of the pink background color. The filter should not have too dense a color, or the color rendition of the blood cells will be degraded. Usually about a CC05G or CC10G filter will suffice.

Mounting Media

After a specimen has been stained, a drop of mounting medium is placed over it and a cover glass pressed into position. The mounting medium may be colored or it may become colored with age. In either case, an effect on color balance will be produced in the recorded image. This effect can be neutralized with a pale CC filter of complementary color.

Canada balsam, for example, tends to become yellow with age through oxidation. The degree of yellowing depends on the thickness of the mounting medium and on its age. A blue CC filter will compensate for this yellow color. However, when a considerable amount of Canada balsam has been used, such as with some whole mounts, and when it has become deeply colored from age, it may be impossible to neutralize the yellow color and still obtain good color balance for the specimen itself. If a specimen slide is placed on a piece of white paper, the presence of a colored mounting medium can be detected before the specimen is photographed. Some mounting media remain colorless and these should be used if possible. The table on page 23 includes a general list of mounting media.

If an existing photomicrograph has

improper overall color because the mounting medium was colored and it is impossible or impractical to reshoot it, a pale CC filter of complementary color can be mounted with the photomicrograph for projection. This technique is practical if the discoloration is not too great.

Chromatic Aberration

A very pronounced undesirable color effect that many people might have difficulty in tracing is a color variance introduced by chromatic aberration in the substage condenser. Both the Abbe and the aplanatic condensers have no correction for chromatic aberration. Even if they are properly adjusted according to the technique of Köhler illumination, an undesirable color effect can be introduced.

When an Abbe condenser has been adjusted correctly to obtain a sharp image of the field diaphragm, that image will usually appear blue at the edge of the diaphragm blades. If the image is not sharp, it will be red, orange, or yellow. Overall coloration may occur also in the specimen image and upset the color balance of the whole photomicrograph. It is therefore necessary not only to focus the image of the field diaphragm sharply with the substage condenser focus but also to gauge this focus by consistently choosing the same color for the slight but inevitable color fringe at the edge of the blades. The aberration effects from these settings and many other factors can be accommodated by suitable filtering. The important thing to do is adopt one or the other color fringe all of the time in order to minimize and fix the aberration factor. Another method is to adjust the substage condenser so that a ring of red is formed just inside the image of the diaphragm blades and a ring of blue is just outside the blades. This will split the difference and help compensate for each color to produce a more neutral balance.

When the edge is blue in one sector but another color in an opposite sector, the lamp filament has not been aligned precisely enough or the condenser is tilted. The steps on pages 30-33 should be repeated carefully. (See Figure 5-7, page 57.)

Even changing microscope slides may affect the focus of the condenser if the new slide has a different thickness. Each time the slide is changed, the opening in the field diaphragm should be decreased and its image examined in the microscope.

Since the deleterious color effects of excessive chromatic aberration may be unpredictable, no suitable filter compensation can be applied. The only suitable compensation is readjustment of the substage condenser to the correct position. To adjust a condenser properly, move it up or down slowly until an image of the field diaphragm, as well as of the specimen image, is visible in the microscope. The edge of field diaphragm image should be sharp and the color adjusted as described above.

The problem of chromatic aberration is minimized when an achromatic condenser is used because it has correction for both chromatic and spherical aberrations. An achromatic condenser is therefore highly recommended for color photomicrography, particularly if apochromatic objectives are used. For best image quality, however, even an achromatic condenser should be focused critically.

Film-Emulsion Variance

The manufacturing of color film involves the coating of light-sensitive emulsions, plus other materials, on a base. The thickness of each emulsion layer, measured in micrometres, must be carefully controlled. If the thickness of any one layer varies from its aim point, an effect on color balance will result. Color-balance variation permitted in manufacture is equivalent to the effect that a CC10 filter would produce on a film of normal balance. Experience has shown that this amount of variation is acceptable to most viewers. However, for critical use, CC filters may be required to produce a desired color balance. A CC10 filter effect is the acceptable tolerance. The direction of variance could be in any one of the six colors: red, green, blue, cyan, yellow, or magenta.

When a photomicrographer changes from one film emulsion number to another, the resulting photomicrograph may show an undesirable background color. The background should be white or very light gray. If it is not, this can be corrected in subsequent exposures by placing a CC10 filter of a complementary color in the light beam.

When many rolls of a particular color film will be used over an extended period of time, several rolls of film with the same emulsion number can be purchased and stored in a refrigerator or freezer. If a deviation from normal color occurs in a film due to manufacturing difference, the amount of deviation can be determined by a filter-balance test with one roll. The test consists of making an exposure with no color compensating filter and an exposure through each of the CC10R, CC10G, CC10B, CC10Y, CC10C, and CC10M filters. When all seven photomicrographs are placed on a 5000 K illuminator, it is likely that one of them will show a clean white or very light gray background. In this way, the correct compensating filter can be determined and used with all subsequent rolls of the same emulsion.

Reciprocity Effect

Reversal color films of daylight type are normally designed for brief exposure times—such as 1/30 second, 1/60 second, and 1/125 second. When the exposing illumination is correct, one of these brief exposure times will usually produce normal color balance in the photomicrograph. Whenever the exposure time for a color reversal film is 1 second or longer, however, an undesirable color effect may be noticeable in the color photomicrograph. With certain color films, even an exposure of 1/8 or 1/4 second may produce a noticeable effect on color balance.

If the light intensity is very low, a long exposure time is often necessary, resulting in a color shift. This color shift is due to the reciprocity effect. To explain: Under normal conditions total exposure equals illuminance multiplied by exposure time ($E = It$). *As the illumination level increases, exposure time must decrease and vice versa, in a reciprocal relationship.* Throughout a normal range of light levels and exposure times, this relationship holds true, but in very low or extremely high light levels, with very short or long exposure times, it may not. The photographic effect will vary with changes in illuminance (I) and time (t). This is especially true for long exposures, which are quite common in photomicrography.

Reciprocity effect is usually apparent as a decrease in emulsion speed at very low light levels. Since a color film contains three emulsion layers, a change in color balance occurs unless all three layers change alike. All three layers, however, may exhibit different reciprocity effects. Color balance can be seriously in error when a very long exposure time is necessary. Long exposure times are often necessary in fluorescence photomicrography, as well as in photomicrography at high magnification with polarized light, or with an interference microscope.

Color shift due to reciprocity effect can often be corrected by placing a single CC10 filter in the light beam before exposure. At very long exposures, greater filter values may be necessary to compensate for color variance.

Recommendations for exposure compensation and for filters to correct color imbalances due to reciprocity effects are usually included in the instruction sheets for photographic films. They are summarized for Kodak films in pamphlets E-1, *Reciprocity Data: KODAK Color Films*, and O-2, *Reciprocity Data: KODAK Professional Black-and-White Films*. (See appendixes of this book.)

Miscellaneous Factors

Several reasons for poor color balance are attributable to improper handling of color film, not to any specific photographic exposure technique or condition. In some cases, the effect is permanent and no filter compensation can be applied. If a photomicrograph shows an undesirable color effect that is not due to any of the factors mentioned previously, one of the following conditions may be the cause.

Improper Film Storage. All photographic films are perishable products that are damaged by high temperature and high relative humidity. Color films are more seriously affected than black-and-white films because heat and moisture usually affect the three emulsion layers to different degrees. For color film, a change in color balance may be accompanied by a change in overall film speed and contrast. None of these effects is entirely predictable. Proper storage of color film is necessary both before and after exposure for consistent color balance. Of course, greater care is necessary under hot and humid conditions. To avoid these effects, follow the manufacturer's recommended storage conditions.

Chemical Fog. Color films not sealed in foil envelopes or snap-top cans should be kept away from any fumes (as from formaldehyde or paraformaldehyde) as well as from any other harmful gases or vapors. Such gases can influence color balance, speed, and contrast; their effects on color film are erratic and unpredictable and may be uncorrectable. Color balance can vary in any direction. Film speed may increase, but contrast will usually decrease—especially with extended exposure to a particular gas—because maximum density is reduced.

Light Fog. Color films should be processed either in total darkness or in lighttight tanks. If a color film has received even a very short exposure to a darkroom light-leak or to a safelight, color balance will be affected. For example, a dark green safelight, which may be used briefly with panchromatic black-and-white films, will cause a green fog on color reversal films.

A loose camera back may allow some light to leak into the camera. The extent of fog in this case depends upon how long the condition exists.

Outdated Film. When possible, color film should be used before the expiration date stamped on the box because a change in color balance is more likely after this date. The magnitude of change depends upon storage conditions during the usable life of that film. If a film has been properly stored under recommended conditions, the rate of change is decreased but not eliminated.

When a quantity of film is stored in sealed containers in a freezer at low temperature (–18 to –23° or 0 to –10°F), the rate of change is greatly decreased. In this case, film can be used beyond the expiration date with reasonable expectation that excessive changes have not occurred. The sealed container of film stored in this way should not be opened for use for at least 2 to 3 hours after removal from the freezer; otherwise, condensation might occur on the emulsion when the cold film is subjected to room temperature.

Radiation Exposure. Whenever a color film has received exposure from a source such as x-rays, radium, cobalt 60, or other radioactive materials, a change in color balance can occur. Such exposure can happen in hospitals, industrial plants, or research laboratories where either radiography or radiotherapy is practiced. If color films are stored or used in areas adjacent to a room where such radiation is present, suitable protection should be provided in the form of lead or concrete shielding of adequate thickness.

Processing Errors. Some changes in color balance in KODAK EKTACHROME Films can be traced directly to a departure from recommended processing conditions. An adequately controlled process should not produce color shifts which amount to more than the equivalent of CC 10 in any direction. The color change can be toward magenta, green, green-yellow, or blue. The causes of such changes can be uneven or insufficient agitation, one or more exhausted solutions, contamination of solutions, improper mixing of chemicals, or incorrect time or temperature. The resulting changes in color balance cannot be compensated adequately. The only remedy is correct processing, as recommended in instructions furnished with processing chemicals.

Viewing Conditions. The conditions under which a transparency is viewed also may have a noticeable effect on its apparent color quality. For critical use, transparencies intended for projection should be judged by projection. If a transparency is to be viewed on an illuminator, such as in an exhibit or other display, the transparency should be judged on an illuminator similar to the one that will be used.

These are very important points and cannot be overemphasized. The practice of holding transparencies up to a window or room light is a haphazard method. The simple transparency viewers that use a 40 W or 60 W light bulb are just as bad. If transparencies are to be viewed on an illuminator, it should be of the graphic arts type, with 5000 K illumination. Projectors with full and reduced illumination options controlled by a selector switch will show two different colors at the two positions. Usually, any two projectors will produce light of different colors unless they are matched by lamp selection or by using CC filters. Beaded screens that are old tend to be yellow because the glue cementing the glass beads to the screen substrate oxidizes. Heat absorbers in projectors may cause small color shifts. Photomicrographic exhibits are often set up in schools, hotels, or public buildings where vast differences in illumination (candle-like light bulbs, fluorescent lighting, or spot lighting) can create tremendous problems in viewing conditions. All of these major and minor sources of color shifts must be taken into account and corrected by the critical photomicrographer.

Color-Balance Corrections

Appearance of Photomicrograph	Possible Cause	Remedy
Slightly Yellow	1. Emulsion variance	Use CC10B filter.
	2. Colored mounting medium	Use blue CC filter (CC10B or more).
	3. Low color temperature of light source	Use light-balancing filter of 82 Series.
	4. Drop in line voltage.	Use constant-voltage transformer.
Slightly Magenta (reddish blue)	Emulsion variance	Use CC10G filter.
Slightly Cyan (bluish green)	1. Emulsion variance	Use CC10R filter.
	2. Heat-absorbing filter in light beam	Use CC10R, or possibly CC20R, filter.
Slightly Blue	1. Emulsion variance	Use CC10Y filter.
	2. Abbe condenser not focused correctly	Adjust condenser for Köhler illumination.
Definitely Blue	1. Ultraviolet radiation present during exposure with arc lamp	Use No. 2B filter to remove ultraviolet radiation.
	2. High color temperature of light source	Use appropriate light-balancing filter of 81 series.
Slightly Green	1. Emulsion variance	Use CC10M filter.
	2. Heat-absorbing filter in light beam	Use CC10M, or possibly CC20M, filter.
	3. Plane-polarized light (one polarizing element)	Use CC05M, CC10M, or CC20M filter.
Slightly Red	1. Emulsion variance	Use CC10C filter.
	2. Abbe condenser not focused correctly	Adjust focus of condenser for Köhler illumination.
Slightly Yellow-Red	Low color temperature of light source	Use light-balancing filter of 82 series.
Definitely Reddish-Yellow	Daylight film with tungsten source—no correction	Use No. 80A filter, plus light-balancing filter of 82 series.

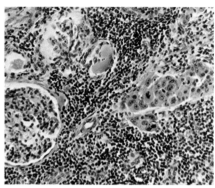

Fig. 5-1:
COLOR BALANCE—*Use of appropriate filters may be required to achieve correct color balance with color films. In the right photo, daylight film was exposed using unfiltered tungsten illumination. The correct filter was inserted into the illuminating beam for the far right photo.*

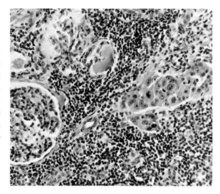

Fig. 5-2
INSTANT FILM BACK—*A motorized KODAK Instant Film Back mounted on the microscope provides rapid access to images.* Photo courtesy of American Optical Corp.

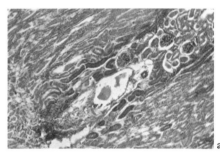

a

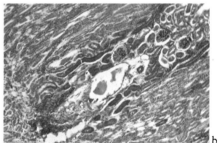

b

Fig. 5-4
CONTRAST INCREASE—
For many microscope subjects, the increase in contrast obtained by exposing KODAK EKTACHROME 50 Professional Film (Tungsten) with an EI of 100 and extending development 30% is very useful. (a) EI 50, normal development. (b) EI 100, 30% increase in development.

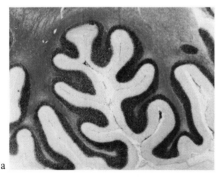

a

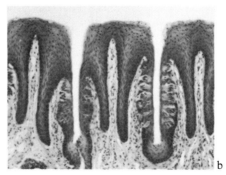

b

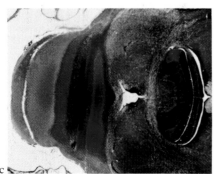

c

Fig. 5-3
INSTANT PRINTS—*These photos (smaller than actual size) are reproduced directly from prints made with KODAK Instant Color Film. (a) Cat cerebellum stained with luxol fast blue counterstained with cresyl violet. (b) Tongue stained with hematoxylin and eosin. (c) Cat brain stem stained with Davenport's silver for nerve fibers.*

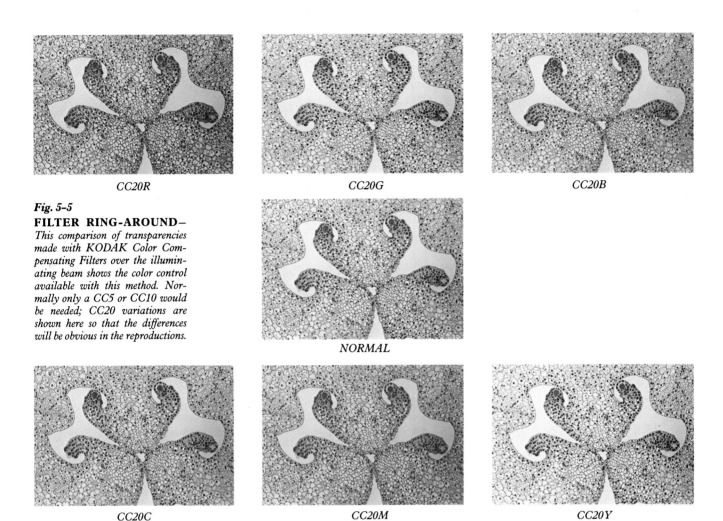

CC20R

CC20G

CC20B

NORMAL

CC20C

CC20M

CC20Y

Fig. 5–5
FILTER RING-AROUND—
This comparison of transparencies made with KODAK Color Compensating Filters over the illuminating beam shows the color control available with this method. Normally only a CC5 or CC10 would be needed; CC20 variations are shown here so that the differences will be obvious in the reproductions.

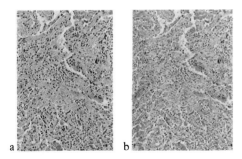

a   b

Fig. 5–6
DIDYMIUM FILTER—*Use of a didymium filter can enhance rendition of certain stains with color films. Here eosin stain is rendered on KODAK EKTACHROME 50 Professional Film (Tungsten) without the didymium filter (a) and with the didymium filter (b).*

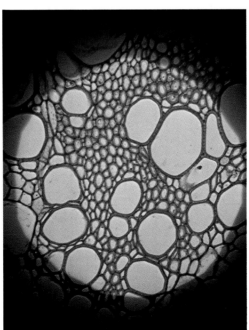

Fig. 5–7
CHROMATIC ABERRA-TION—*Incorrect condenser adjustment or misaligned lamp filament can cause color balance errors. Here the blue fringing at one side of the field diaphragm image indicates that the lamp is not aligned or that the condenser is tipped.*

Chapter Six

BLACK-AND-WHITE PHOTOMICROGRAPHY

Black-and-white photographic materials are often used in photomicrography for a variety of reasons. Illustrations in scientific books, journals, and reports are usually in black-and-white because of the additional expense involved in printing color. Although color might be more desirable, a good photomicrograph in black-and-white is always acceptable. Black-and-white materials have the advantages over color of better contrast control, better UV response, less critical exposure, and quick results.

Black-and-White Films

A black-and-white photomicrograph is almost always recorded as a *negative*, which must be printed onto a suitable print material. Reversal materials are seldom used. The selection of a negative material, usually a film, requires knowledge of photographic characteristics such as color sensitivity, contrast, granularity, resolving power, speed, and exposure and development latitude. *(The appendixes contain more information on film selection.)*

Color sensitivity is an inherent characteristic related to the response of a film to colors of the spectrum. A black-and-white film may be *blue-sensitive, orthochromatic, panchromatic* or *infrared-sensitive.* Blue-sensitive materials respond only to blue light, and to ultraviolet radiation; orthochromatic materials have sensitivity extended into the green; and panchromatic materials are sensitive to all visible colors. Infrared films record all the visible light plus infrared radiation. Panchromatic films (or plates) are most often used in black-and-white photomicrography, since stained, colored specimens are common. Color sensitivity is a fixed property of photographic material; it is not subject to change by alteration in processing as are other characteristics.

Generally, a film with very fine grain, high resolving power, and moderate contrast is best for black-and-white photomicrography. Film speed is of less importance unless the subject is in motion and a film of high photographic speed is needed. In this case, some sacrifice in graininess, contrast, or resolving power

may be necessary in order to record a satisfactory image.

In selecting a suitable material, consider the specimen to be photographed. If the image exhibits relatively low contrast, a high-contrast film may be needed. Conversely, if the image has high contrast, a low- to medium-contrast film may give best rendition of detail. If the negative must be enlarged, a film that has extremely fine grain can be an advantage.

Most films are classified as having low, medium, or high contrast. While this property can be altered to some extent by changing development time in a given developer or by changing the developer itself, the best practice is to follow the manufacturer's recommendation as to developer and development time. Deviation from an established procedure may have a pronounced effect on graininess, resolving power, or speed. When the experienced photomicrographer does make a change, the change is usually in the direction of extending the development to increase film contrast. An increase of 20, 30, or even 50 percent in development may be necessary to achieve the necessary contrast increase. A versatile black and white film that can be processed in different developers to a wide range of contrasts is KODAK Technical Pan Film 2415 (see Appendix).

Resolving Power and Graininess

Resolving power refers to the ability of a photographic material to record fine details distinguishably. It is expressed as the number of line pairs per millimetre that are recognizable as separate lines and spaces in a photograph. Resolving power is determined for a particular film by photographing—at greatly reduced size—a parallel-line test chart with a high-quality lens. The image is then examined through a microscope to determine the number of lines that can be resolved. Determination of resolution depends on the test-object contrast; comparison of different films is made only with test objects of equal contrast.

Graininess is related not only to the size of silver grains produced in a film after development but also to the irregular distribution of silver grains in the emulsion. The degree of

graininess will vary, depending on the type of developer used and, somewhat, on development time. A fine-grain developer may produce less graininess, but speed and contrast will be reduced. Generally, fast films have coarser grain than slow films.

The graininess designation for a photographic material is usually an indication of inherent resolving power. A film with very fine grain is thus *capable* of high resolution. Remember, however, that resolution in photomicrography is *not* dependent on the resolving power or on the graininess of a photographic material. High resolution in the image can be achieved only by a quality microscope used efficiently. You should not try to substitute photographic enhancement for good microscope technique. While it has been argued that very high-contrast film techniques can effectively alter the constant in the resolution equation, these techniques have limited use and cannot substitute for adequate preparation of the microscope image for photomicrography.

Detail may be made more visible by photographic enlargement from a negative, however, if the recorded detail in the microscope image was greater than that visible to the eye (about 10 lines per millimetre). In this case, the detail or resolution of detail that the eye can see is improved but certainly not over what was in the microscope image. For instance at 50X, 2000 lines per millimetre in the specimen becomes 40 lines per millimetre at the microscope eyepiece. Since the eye can resolve only about 10 lines per millimetre, 4X enlargement will show greater detail on the print than was visually evident but still not more than was already resolved by the microscope.

It is a common misconception that a film with very fine grain is necessary to achieve high resolution in photomicrography. The real advantage of using a film with very fine grain is the probability of good image quality. Even though the recorded image cannot be better than the microscope image, the photographic material should not degrade image quality. A film with high resolving power will show an image more effectively than one of coarse grain or one with low resolving power. For example, a microscopic specimen contains fine details, whose separation is equivalent to about 2000 lines per millimetre.

The microscope optics in use may resolve details at a magnification of 500. The separation of details in the image is now equivalent to 2000/500, or *only four lines per millimetre*. Any film is capable of such low resolving power, but a fine-grain film will demonstrate actual microscope resolution more clearly.

Selection of a fine-grain film will also allow some photographic enlargement without lowering image quality. Too much enlargement, however, may result in empty magnification; that is, although the image is larger, no more details are resolved. This possibility must be considered when small-size films, such as 35 mm, are used in photomicrography. When an image is recorded on a small film area, further enlargement is necessary in order to distinguish details in a print. However, if maximum usable magnification has been achieved in the microscope and recorded on the film, no further enlargement is possible without degrading image quality. When enlargement is contemplated, the final viewing magnification should be considered to determine the required resolution on the film.

As an example, suppose that an image is recorded on film at 500X but that the microscope optics in use are capable of 1000X. The recorded image could still be enlarged 2X without causing empty magnification. If the same image were recorded at only 250X, then a 4X photographic enlargement would be acceptable. If the image were recorded at 1000X, however, little further enlargement should be done. These facts are of particular interest and importance when photomicrographs are made for publication.

Since the magnification in the microscope is computed for the virtual image at 250 mm (10 inches), the above comparisons apply to a glossy print viewed at 250 mm (10 inches). Were the negative enlarged twice, the same detail would be observable at 20 inches, but no new detail would be apparent at 10 inches. Another factor governing the initial magnification is the NA of the system—a suitable objective for resolving the structures must be chosen and magnification at the film plane controlled with a good eyepiece.

Exposure Latitude

Exposure latitude is the range of exposures from underexposure to overexposure that will produce acceptable results. With negative films, more latitude is usually possible toward overexposure. When in doubt as to correct exposure, then, it is safer to overexpose a negative film slightly. The amount of acceptable overexposure, however, is limited by

increased graininess and difficulty in printing. Ideally, of course, the optimum exposure will produce the best negative. With transparency films, there is usually slightly more exposure latitude on the underexposure side.

Generally, high-contrast films provide the *least* exposure latitude, and low-contrast films, the most. Also, latitude will vary, within limits, according to selection of developer and development time. As development time increases, exposure latitude as seen in the processed negative decreases because of the increase in contrast.

Development Latitude

A negative film (or plate) capable of a wide range of contrasts—by manipulation of developer or development time—is highly desirable in photomicrography. Such a material can be used in a great number of applications. Very few materials, however, have this capability. A film designated as low in contrast can seldom be used to produce high contrast; one designated as high contrast should be used cautiously for low-contrast results. In both cases, undesirable effects on graininess, speed, and tonal rendition may result. A film of medium or moderate contrast will usually be capable of the widest development latitude; a notable exception is KODAK Technical Pan Film 2415 which has inherent high contrast but which can be processed to intermediate or normal contrast.

Film Speed

In the U.S., film speeds are arithmetic; the number assigned to a film (ASA) is directly related to its sensitivity. A high-speed film of ASA 400 is twice as fast as one rated at 200, for example. Speed ratings are given on instruction sheets supplied with the different films and apply for daylight or tungsten illumination as indicated.*

The rated speed for a film given in the instruction sheet only applies when the film is developed according to recommendations. Deviation from the recommended development procedure may have a pronounced effect on speed and contrast. In such cases, the microscopist must experiment to establish a useful exposure index.

*Now appearing is the designation ISO (International Standards Organization) film speed. This expresses speed both arithmetically (same number as ASA speed) and logarithmically (same as DIN speed).

Filters in Black-and-White Photomicrography

In black-and-white photomicrography, filters are used primarily for control of image contrast. An increase in contrast is often desired in order to make a specimen stand out against the background or to differentiate between colored elements, which may appear to have equal brightness on black-and-white film. In the latter case, a filter might be employed to absorb one color more than the other. Otherwise, the colors could record as equal gray tones, particularly if the panchromatic film used had equal sensitivity to both colors.

When a specimen color is a very pale one against a bright background, it may record as a pale gray against a white background. A filter that absorbs this specimen color efficiently will render it as a darker gray; increased image contrast will result. Maximum contrast occurs when the specimen color is completely absorbed; intermediate contrast, when it is partially absorbed; and minimum contrast, when the filter color is the same as, or similar to, the color in the specimen. (See Figure 6-1, page 62.)

KODAK WRATTEN Filters

KODAK WRATTEN Filters are thin (0.1 mm) films of gelatin mixed with dyes. For durability, they are overcoated with a thin lacquer layer. Gelatin film squares can be cut with scissors to fit specific equipment. Plain gelatin filters are appropriate for experimental or temporary use. Such filters, however, are subject to abrasion, accumulate grease and dirt in handling, and collect dust from the air. For permanence they may be mounted in glass in the appropriate size for insertion in a microscope light beam. In either form they should be protected from heat with a heat-absorbing filter. Continued use without protection may cause fading. Glass filters can be cleaned easily, in the same way lenses are cleaned. Gelatin filters are *not* easily cleaned.

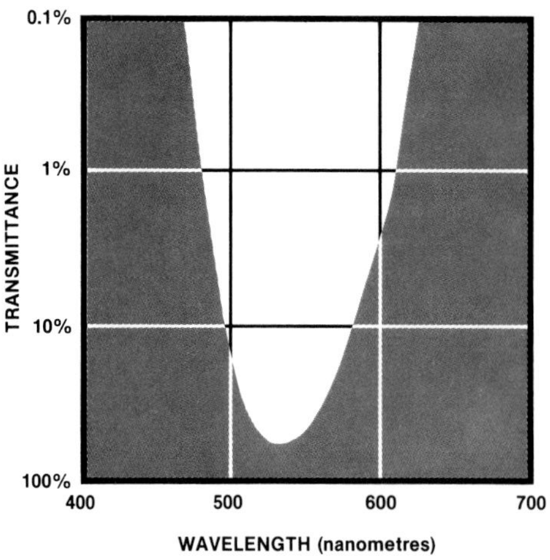

0.1%

TRANSMITTANCE

100%
400   500   600   700

WAVELENGTH (nanometres)

Fig. 6–2

TRANSMITTANCE CURVE–
*KODAK WRATTEN Gelatin
Filter No. 58 (green).*

ABSORPTION (RELATIVE)

400   500   600   700

WAVELENGTH (nanometres)

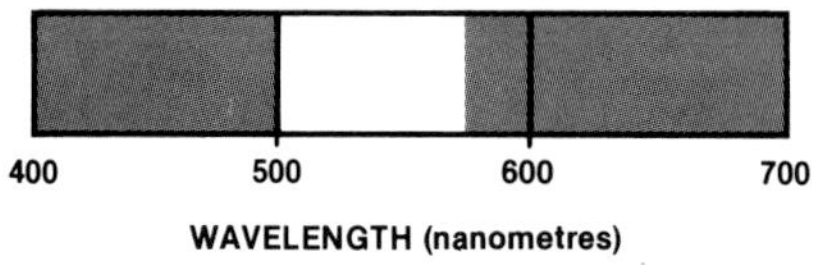

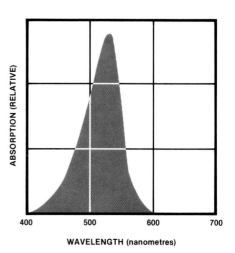

Fig. 6–3

BAR CHART—*Simplified transmittance data (above 10 percent shown as clear area) for a KODAK WRATTEN Gelatin Filter No. 58 (green).*

400   500   600   700

WAVELENGTH (nanometres)

Fig. 6–4

ABSORPTION CURVE—*Safranin O. Absorption range (above 10 percent) is 470-550 nm.*

Filter Transmittance Curves

Both the absorption and transmission characteristics of all KODAK WRATTEN Filters are given in *KODAK Filters for Scientific and Technical Uses*, KODAK Publication No. B-3. A typical curve for one filter, the No. 58 (green), is shown in Figure 6-2. The white area indicates the visible light *transmitted* by the filter; the colored area shows what regions of the visible spectrum are *absorbed* by the filter. The numerical values of the exact percentage of transmittance at specific wavelengths are given in the data book. The dominant wavelengths are also given to indicate maximum transmittance for a variety of sources.

For practical purposes, the visible spectrum can be divided into three primary color regions. The range from 400 to 500 nm appears visually as blue, from 500 to 600 as green, and from 600 to 700 as red. A filter that transmits all or most of one region will appear to have the indicated color. The No. 29 filter, for example, transmits all of the red region and appears red. Most of the light transmitted by a No. 61 filter is between 500 and 600 nm. This is a green filter. The No. 47 filter transmits most of the region between 400 and 500 nm, and is blue. Other filters transmit greater or less portions of these regions. The color of a particular filter is usually due to the greater part of the visual spectrum transmitted by it. Some filters will transmit all or most of two spectral regions and absorb the third. A No.

12 filter, for example, is yellow but will transmit both the green and red ranges from 500 to 700 nm. Since this filter absorbs all of the blue region, it is often called minus blue. A No. 32 magenta filter is often called minus green, since it absorbs green but transmits red and blue quite freely. A No. 64 deep blue-green (cyan) filter transmits most of the blue and green regions and absorbs all red.

A simplified method of showing transmittance and absorption areas of the spectrum is to make a bar chart. Such a chart will provide a simple index to the ability of a particular filter, or combination of filters, to do a certain job of color transmittance or absorption. A bar chart for a filter is constructed from its transmittance curve. (An example is shown in Figure 6-3.) The *white* area shows transmittance greater than 10 percent.

Bar charts for several filters and filter combinations are shown in Figure 6-5. In these charts the transmittance of 10 percent or more is shown.

It is not always possible to recommend a specific filter, even when stain colors or other specimen colors are known. Visual inspection of a specimen through different filters is usually the best guide when selecting a filter for photomicrography. Experienced photomicrographers can examine a specimen and often select the best filter with little trouble, since they are usually familiar with stains and filters and with their effects in a photomicrograph.

Biological Stains

The elements of a microscopic specimen, whether a tissue section or a smear, usually have little natural color; nor do they absorb light to any great extent. In order to make such elements visible, they are often colored with biological stains or dyes. Two or more such stains are sometimes used to color specific elements differently, producing a color contrast between them. The nucleus of a cell, for example, may be stained one color, and the cytoplasm, another. A single element may be stained to produce contrast between it and the background. In a brightfield microscope the background normally appears white or very light gray, so a stained object will appear colored against a very light background.

The colors of stains used for this purpose vary considerably. In the range of biological stains commonly used, all colors are represented. Although many stains appear to have a similar color, their absorption bands occur in slightly different portions of the spectrum. Many stains, for example, have a reddish color and may appear identical visually. The difference between them is only evident by analysis of the wavelengths absorbed by each. This analysis can be done with a spectrophotometer, which produces an *absorption curve*. Such curves have been published for many common biological stains. The absorption curve for a given stain indicates the position

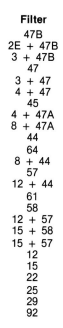

Filter

47B
2E + 47B
3 + 47B
47
3 + 47
4 + 47
45
4 + 47A
8 + 47A
44
64
8 + 44
57
12 + 44
61
58
12 + 57
15 + 58
15 + 57
12
15
22
25
29
92

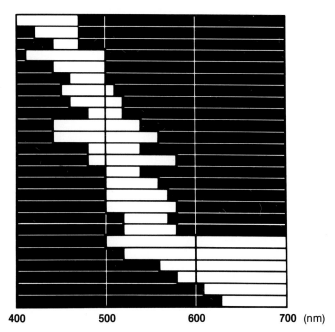

400   500   600   700 (nm)

Fig. 6-5
TRANSMITTANCE OF FILTERS—*The transmittance (over 10 percent) for common KODAK WRATTEN Gelatin Filters and useful combinations.*

Stain

Acid Fuchsin
Aniline Blue
Azure C
Basic Fuchsin
Brilliant Cresyl Blue
Carmine
Congo Red
Crystal Violet
Darrow Red
Eosin Y
Erythrosin B
Ethyl Eosin
Light Green SF
Methyl Green
Methylene Blue
Neutral Red
Phloxine B
Orange G
Safranin O
Sudan IV
Tartrazine
Toluidine Blue

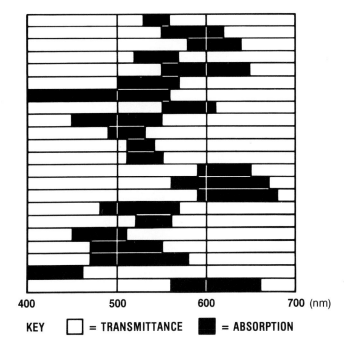

400   500   600   700 (nm)

KEY   ☐ = TRANSMITTANCE   ■ = ABSORPTION

Fig. 6-6
ABSORPTION OF STAINS—*The approximate absorption (over 10 percent) of some biological stains. (Compare with filter transmittance.)*

of its absorption band in the visible spectrum. See Figure 6-4.

In black-and-white photomicrography of stained specimens it is often necessary to select a filter to produce contrast between a colored specimen and the background. Knowledge of the absorption characteristics of particular stains is helpful in selecting an appropriate filter. Absorption characteristics of stains can be obtained either by spectrophotometry or by reference to pertinent literature where spectral absorption curves are published. A simplified *bar chart* can be made from an absorption curve (as was done with filters). Compare the bar charts of filter transmittance (Figure 6-5) and stain absorption (Figure 6-6).

Increasing Contrast

Filters are often used to increase contrast in photomicrography of blood smears. In the United States, blood smears are commonly stained with Wright's stain, a combination of methylene-blue and eosin stains. The first stain usually appears deep blue, and the second, light red. A green filter, such as a KODAK WRATTEN Gelatin Filter No. 58, will show more absorption for the eosin color and will render it with good contrast against the background. Another common stain combination, hematoxylin and eosin, usually requires a green filter also. The particular green filter depends on stain concentrations. The No. 58 filter applies if the eosin color is pale, a lighter yellowish-green filter, such as a No. 11 or No. 13, when the color is more intense.

When a specimen color is moderately dense, detail within the specimen can be recorded if the filter color is the same as, or similar to, the specimen color. (See Figure 6-7, page 62.)

If maximum contrast is desired between a colored specimen and the background, the filter should *absorb* the specimen color completely. In this case, the filter color is *complementary* to the specimen color.

One technique that can be used for selecting a filter is to examine the specimen through the microscope while trying different filters in the light beam. When the desired contrast or detail rendition is achieved, make a photomicrograph through that filter on a panchromatic photographic material.

When contrast between specimen and background is desired, use the filter circle (Figure 6-8, page 64) as an aid in choosing contrast filter colors.

When selecting a filter for contrast, you must be careful not to increase contrast *too much* in a photomicrograph. The result will

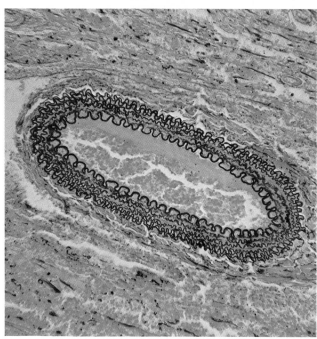

Fig. 6-1
EFFECT OF FILTERS—*Compare the separation of tones of the blue and green stains (a) when rendered in black-and-white through KODAK WRATTEN Gelatin Filters. (b) No. 25 (red), (c) No. 58 (green), (d) No. 47B (blue).*

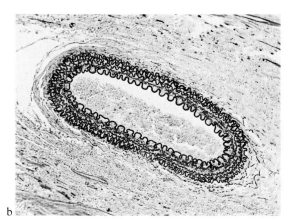

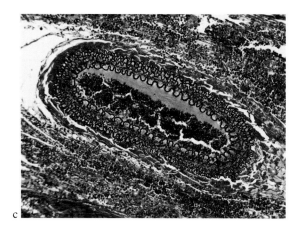

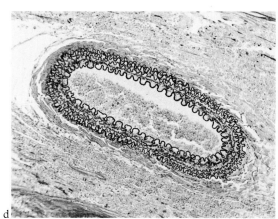

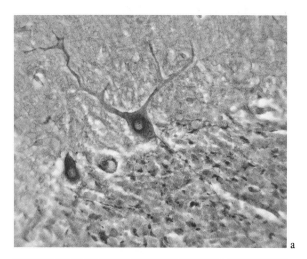

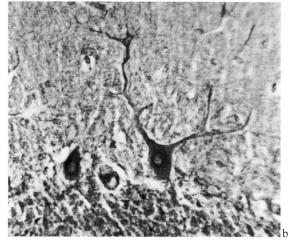

Fig. 6-7
TONE RENDERING—*Selection of filter influences tone and contrast of black-and-white rendering. The predominant yellow of silver impregnated nerve cells (a) required the use of a KODAK WRATTEN Gelatin Filter No. 11 (light yellow-green) to lighten the background and show detail in cells.*

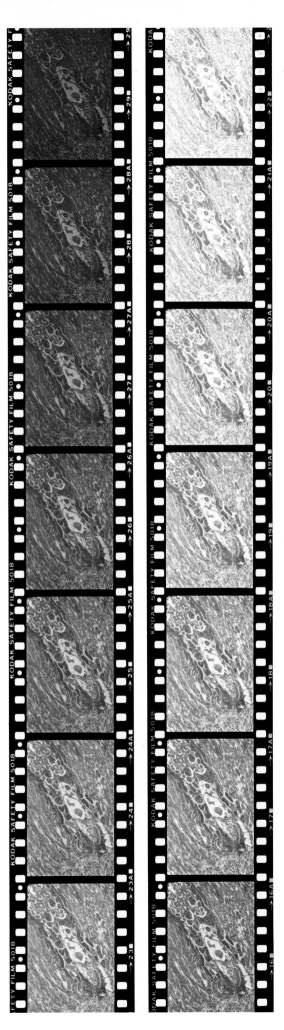

Fig. 7-1
EXPOSURE SERIES—*In an exposure series on 35 mm color transparency film, exposure was varied over a range of 4 stops (16X) at ⅓ stop intervals. The ⅓-stop change between frames was made by using 0.1 and 0.2 neutral density filters to reduce intensity of this illuminating beam between the full stop changes provided by change of exposure time (shutter speeds).*

be a blocking of dark areas and a loss of fine detail. The use of high-contrast film is not necessary if sufficient visual contrast is obtained through the filter.

Since stain colors and their absorptions vary considerably, final filter choice will depend upon the specimen. For example, a blue-appearing stain, such as aniline blue, may transmit some red as well as blue. A red filter would not completely absorb this color, but would show some transmittance. The color of methylene blue, on the other hand, is completely absorbed by a red filter such as a No. 25. It would be obvious from either the absorption curve or a bar chart for aniline blue that a red filter would not do a complete job of absorption if a panchromatic film were used to record the image. A deep green filter would be better. Try a combination of a No. 22 and a No. 58 filter. Check the resulting change in contrast by visual inspection in the microscope.

Panchromatic materials differ somewhat in their response to red, green, and blue light. In general, with tungsten illumination a yellowish green filter, such as a KODAK WRATTEN Gelatin Filter No. 11, should be used for correct gray-tone rendering of multi-colored specimens. If the illumination is of daylight quality (such as with a xenon arc), use a yellow filter, such as a No. 8 filter, because of the higher proportion of blue in daylight illumination. If a film has a red sensitivity that is higher than normal, try a No. 13 filter for tungsten illumination and a No. 11 filter for illumination of daylight quality.

It is possible that in some instances none of the above filter recommendations will produce adequate contrast between elements of a specimen and the background. For instance, two colors having similar brightness values may be either transmitted freely or absorbed equally by either the No. 11 or No. 13 filter. In this case, a filter must be found that will absorb one color fairly well and partially absorb the other, so that both will stand out against the background when photographed.

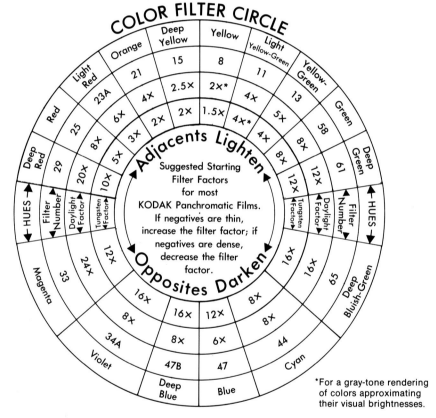

Note: If conditions require long time exposures, corrections for reciprocity effect in addition to the corrections for the filter factor may be necessary.

*For a gray-tone rendering of colors approximating their visual brightnesses.

Fig. 6–8
COLOR FILTER CIRCLE

Chapter Seven
DETERMINING EXPOSURE

As in other forms of photography, exposure in photomicrography is the result of light acting upon a sensitized film. Exposure is influenced by light intensity (image brightness) and by exposure time. In order to record an image of a particular brightness, a specific exposure time is necessary, depending upon the film speed.

Exposure time can be determined either by making an exposure test series or by using a sensitive exposure meter. Because image brightness in photomicrography can be quite low, particularly at high magnification, an exposure meter should be sensitive to very low light levels. When making an exposure test series, be sure to place any necessary filters in the light beam, whether exposing a color film or a black-and-white film.

Reciprocity Effect

All photographic emulsions are subject to an effect described as reciprocity-law failure. The reciprocity law states that the product of the intensity of the illumination falling on a film and the exposure time equals the amount of exposure ($E = It$). This law does apply to most black-and-white films for exposure times of about 1/5 to 1/1000 second and to color films over a somewhat narrower range. In other words, an exposure of 1/60 second at $f/11$ is equivalent to an exposure of 1/30 second at $f/16$.

For times outside the normal range, the effect of reciprocity-law failure can be seen as underexposure or a change in contrast in black-and-white films and underexposure or a change in color in color films. The word failure, in this connection, does not imply a shortcoming of the film, but merely that the reciprocity law does not hold for very short or very long exposure times.

In other words, the effective sensitivity of a film emulsion varies linearly with illumination level and exposure time over a normal range of values. Each emulsion has its greatest response within a particular range of illumination values. On either side of this range the response decreases. Extra exposure is needed to obtain normal density and contrast in black-and-white films. Since color films have three color-sensitive layers, each of which may be affected differently by

the reciprocity effect, both exposure and color-balance correction may be necessary.

Low-Intensity Reciprocity Effect

In photomicrography, it is more likely that relatively low intensity of illumination will occur and long exposures will be required. This can lead to the low-intensity reciprocity effect.

In black-and-white films, the normally calculated exposure for low intensities will result in negatives that lack shadow detail. If, in order to correct for this, the exposure time is lengthened still more, the density in the highlight areas becomes disproportionately greater than that in the shadows, resulting in an increase in effective contrast. The reciprocity effect is greater in areas of the specimen where the illumination is relatively low. This can be compensated for by decreasing the amount of development.

Color films will usually be underexposed and show a color shift with long exposure times. Since each color film reacts somewhat differently, corrections cannot be generalized. The information is included in film instruction sheets or pamphlets either as a suggested exposure adjustment or as an effective film speed along with a recommended color correction filter.

Exposure Tests

Roll Film

When exposing a 35 mm color transparency film or other size of roll color film, test by exposing several frames, each for a different length of time. Keep image brightness constant, of course. Process the exposed film and examine the individual frames on a standard 5000 K illuminator to determine which one received the correct exposure. The best transparency is the one that exhibits detail throughout, in both shadows and highlight. (See Figure 7-1, page 63.)

A convenient series includes all of the exposure times available with the camera shutter. These may range from 1 second to 1/125 second or 1/250 second. If the shutter has a *time* (T) setting, a few long exposures

can also be made (2, 4, and 8 seconds, for example). Out of this wide range of exposure times, one should be correct for the particular film and the existing image brightness.

An exposure series can be done by changing the intensity of the illumination, but this should not be done by changing the voltage supplied to the lamp since this will change the color quality of the illumination. Rather, the lamp voltage should be held constant and neutral density filters introduced in the illuminating beam with each successive exposure. Introduction of 0.1 density in the light beam represents reduction of the intensity by 1/3 stop. A density of 0.3 will reduce the intensity by a full stop (to 50 percent of the original value). For an initial trial then, reduce intensity by introducing 0.3, 0.6, 0.9, . . . densities for a whole-stop exposure series.

Even when using color film, it may be convenient to make the exposure series by substituting a roll of black-and-white film, processing it, and evaluating the negatives. Correlate the film speed of the black-and-white film with any other film that might be used. As an example, suppose the test film has a speed of ASA 32 and a new film has a speed of ASA 125. Suppose also that the exposure test with the ASA 32 film indicates 1/30 second to be the correct exposure time. Since the ASA 125 film has about *four times* the speed of the ASA 32 film, it will require only *one-fourth* the exposure time, or 1/125 second.

Sheet Film

When a camera accepts sheet films, a series of test exposures can be made on one film sheet by the following procedure:
1. Pull out the dark slide of the film holder until the entire sheet is uncovered in the camera. Make an exposure for one unit of time—for example, 1 second, 1/2 second, 1/125 second, or whatever initial time may be indicated by the image brightness level.
2. Push in the slide about an inch and *repeat* the exposure for the same unit of time.
3. Push in the slide another inch and give an exposure for two units of time.
4. Continue to push in the slide to cover approximately 1-inch steps, exposing each step for twice as long as the previous one. The successive steps will then have

received exposures of 1, 2, 4, and 8 times the initial time unit.

Develop, fix, and wash the film as recommended in the instruction sheet. Examine the resulting negative to select the step that shows the best reproduction of the subject. (See Figure 7-2.)

Exposure Calculation

Once the correct exposure time has been found for a particular set of optical conditions, the correct times for new conditions *with the same microscope and illumination* can be calculated. The factors that affect image brightness are changes in magnification or in numerical aperture. Magnification, of course, will change if a different eyepiece, a different objective, or a new eyepiece-to-film distance is involved. Also, if the objective is changed to one of higher or lower magnifying power, the numerical aperture will be different.

If the optical conditions for which the exposure time was originally determined by test are called *standard*, they can be compared mathematically with the *new* conditions to calculate the *new* exposure time, according to the following equation:

$$\frac{\text{New Exposure Time}}{\text{Standard Time}} = \left(\frac{\text{Standard N.A.}}{\text{New N.A.}}\right)^2 \times \left(\frac{\text{New Magnification}}{\text{Standard Magnification}}\right)$$

The application of this equation is illustrated by the following example: The correct exposure for KODAK EKTACHROME 64 Film was determined to be 1/125 second (0.008 second). Suppose that a 10X eyepiece and a 20X objective (NA = 0.50) were used. Magnification is 200X. These are the standard conditions. With the same film, a new exposure will be made at 430X, using the same 10X eyepiece and a 43X objective of NA = 0.65. What is the new exposure time under these conditions?

$$\frac{\text{New Exposure}}{0.008} = \left(\frac{0.50}{0.65}\right)^2 \times \left(\frac{430}{200}\right)^2$$

New Exposure = 0.022 second or about 1/45 second

In this way exposure time can be calculated when any of the optical exposure factors are changed. It is even possible to set up a table to include all exposure times for a color film

Fig. 7–2
STEPPED EXPOSURES– *The test exposures on sheet film were stepped off in a series (as described in the text) representing 1, 2, 4, and 8 times the basic exposure value.*

used under different conditions. This would require, of course, that the microscope and the illumination be correctly adjusted for each condition.

In black-and-white photomicrography, specific light filters are often used for contrast control. When different filters are used, as they often are, their filter factors must be considered in working out exposure determinations unless you use an integral exposure meter.

When films of different speed are used, similar correction calculations can be made by applying the following equation:

$$\frac{\text{New Exposure Time}}{\text{Standard Time}} = \frac{\text{Standard Exposure Index}}{\text{New Exposure Index}}$$

The application of this equation is illustrated by the following example: The correct exposure for KODAK PANATOMIC-X Film with a speed of ASA 32 is 1/10 second (0.1 second). What will be the exposure time for KODAK EKTACHROME 160 Film (Tungsten) with a speed of ASA 160?

$$\frac{\text{New Exposure}}{0.1} = \frac{32}{160}$$

New Exposure = 0.016 second = 1/60 second

In this way, exposure time can be calculated when any film speed is changed for another.

Judging Exposures

A good photomicrographic exposure should always show the subject to best advantage, whether the image is on a reversal color film or on a negative material. It is easier, however, to judge correct exposure on a reversal material since this type of film has very narrow *exposure latitude*. If an exposure time is not right, the fact is immediately apparent. Overexposure will wash out the light areas, with a loss of *highlight* detail. Underexposure will cause a darkened appearance overall, with loss of detail in the darkest, or *shadow*, areas. A good exposure on a reversal film will usually show detail in all areas where detail was evident in the specimen. A well exposed transparency has some very slight (0.1) density in highlight areas.

Negative materials, both color and black-and-white, exhibit much more exposure latitude than do reversal color films. Therefore, judgment of the best exposure time is more difficult. In general, exposure should be controlled to provide best rendition of detail in the *darker* areas of the subject. These areas will be represented on a negative by light areas. If the film is underexposed, very little detail will be evident in these areas of the negative. When such a negative is printed, the background and highlight areas may print very well, but the darker areas will be too dark, with few, if any, details. With negative materials, more overexposure than underexposure can be tolerated. However, considerable overexposure of a negative may cause loss of detail in both the highlight areas and the darkest areas.

Filter Factors

When a colored filter is placed in the illumination beam in black-and-white photomicrography, it will *absorb* a certain amount of light. The amount of light absorbed depends upon the particular filter. One may absorb more, or less, than another. If the exposure time for a black-and-white film is determined with *white light* (without any filter), it must be adjusted (increased) when a filter is used since the image brightness will be decreased. The amount of increase in exposure time is known as the *filter factor*. The exposure time without the filter is multiplied by the filter factor to determine the exposure with the filter in place. Filter factors for Kodak black-and-white films are published in the instruction sheets packaged with the films. Depending upon the filter and whether or not you want to drop out a color, you may want to use a smaller filter factor, or none at all.

When exposure time is determined for a black-and-white film by making exposure tests, the filter can be in position. Correct exposure will then be achieved without resorting to the filter factor. If a new exposure of a different subject is to be made, and a different filter is necessary, the new exposure *must* consider the change in filter factors. For example, suppose that an exposure was made on a black-and-white film with a filter having a factor of 4. The new exposure requires a different filter whose factor is only 2. The new exposure time will be *one-half* the previous one. Or, if the new factor were 8, the new exposure time would be *twice* the original exposure.

If exposure is determined with an exposure meter, only white light (unfiltered) should be used for black-and-white films. The photocell in an exposure meter does not have a uniform response to all spectral colors, so measurement of exposure with the filter in place could be erroneous. Thus, the best practice is to measure the exposure time for white light and then apply the filter factor for finding the correct exposure.

If color films are to be exposed, filter factors do not apply with through-the-lens meters. Light balancing filters should already be in place for any exposure determinations; see the film instruction sheet for required alteration in film speed.

Determining Filter Factors

Because of the difference in spectral response of various photographic materials and the difference in color quality of different types of illumination, published filter factors can only be considered as approximate. If an exact filter factor is necessary for specific conditions, it must be determined experimentally. This can be accomplished by making a step-exposure series on the film to be used, with the illumination to be used, and, of course, with white light. It is not necessary to have a specimen in place on the microscope stage for the exposure series—the background brightness will suffice. The exact series of exposures will be governed by the actual brightness level. If the level is low, a *power of 2 series*—such as 1/2, 1, 2, 4, 8, and 16 seconds—could be used. Shorter exposure times would apply for a high brightness level. Microscope adjustments must not be changed during the test.

If a corresponding step-exposure series were then made with a filter in place, both series could be compared to determine which steps matched for density. If no two steps matched, either another series with closer steps could be made or the filter factor could be determined by interpolating a value between the two density steps. This requires a knowledge of densitometry, but is the most exact technique.

Suppose, for example, that 2 seconds of exposure produced a medium density in a series made with white light. When the filter was placed in the light beam and another exposure series was made, it took 8 seconds to produce a density equal to the 2-second exposure in the first series. The filter factor would then be 8 divided by 2, or 4.

Experimental determination of filter factor is especially useful when filters are used in combination, since factors are seldom published for combinations. Once the combination filter factor has been determined, it can then be applied in all future exposures, provided that the same film and illumination are used.

Exposure Meters

Some types of exposure meters are made specifically for photomicrography and are sensitive enough to respond to light through a wide range, from very low to very high brightness. A meter scale may be precalibrated by the manufacturer to give a direct reading of exposure time, with various settings for film speed. Obviously this is the most convenient type of meter. Other meters may be very sensitive to light but have readings in terms of units of illuminance, such as footcandles. In this case, the manufacturer often provides a simple calculator, table, or graph so that brightness readings can be converted to exposure times.

If a meter reads brightness but no device is provided for correlating brightness and exposure time, the meter must be calibrated. This can be done by making a brightness reading of the image at low magnification (about 100X). The brightness reading is recorded, as indicated on the meter scale, for later reference. Then a series of exposures is made at various shutter speeds on a reversal color film. After the film is processed, the correct exposure time is selected. A table (or a graph) can then be made that includes the brightness reading, the correct exposure time for that reading, and the speed of the color film exposed. If another reading were made at a different magnification, one could use the existing data to find the correct exposure time. For example, if the new reading at lower power indicates *twice* the brightness, then the new exposure time will be *one-half* the previous one; if a new reading at higher power is only one-half the first one, the new exposure time will be *twice* as much. Exposure time for a given film will vary *inversely* as the brightness of the image increases.

This system could be used for any light meter not calibrated in terms of exposure time or film speed, including meters normally intended for conventional photography outdoors and indoors.

Probe Meter

One type of meter is equipped with a probe containing a photocell. The probe is placed in the microscope eyepiece tube, with the eyepiece removed, in order to read image brightness. When a probe type of meter is used in this way, the *same* eyepiece should

always be used in the microscope for photography. Otherwise, the meter readings will not always be valid. This method is seldom necessary today.

When an eyepiece camera with a beam splitter is used over the microscope, light readings can be made from the light emitted from the eyepiece of the beam splitter. However, image brightness here will often be *less* than that seen by the film because the division of light varies with different beam splitters. In some systems, 90 percent of the light goes to the camera and only 10 percent is seen visually through the observation eyepiece. The actual division, however, may be 80-20, 70-30, or even 50-50, depending on the specific eyepiece camera in use. Hence, the amount of light division will influence the calibration of a light meter. Actually, it does not matter what the division of light is as a result of the beam splitter, as long as the light is read in the same place and in the same way as in the initial test roll.

Some of the probes today are quite small, and at least one type can be used on the screen of a removable ground-glass viewfinder of an attachable 35 mm single-lens-reflex camera in the same way that a larger probe would be used to take readings from a large-format ground glass.

Making Exposure Readings

Exposure readings can be made in various positions—at the film plane of the camera, anywhere between the eyepoint of the ocular and the film plane, in the eyepiece tube of the microscope with the eyepiece removed, or from the observation eyepiece of a beam splitter. The best position, of course, is at the film plane since the image brightness there is the same as would be recorded on the film. This position, however, is not always accessible because the camera must be closed during photography. The ground-glass screen of a sheet-film camera allows this type of reading before insertion of the film holder. For test purposes, use a slide that has an average amount of material in the beam—not excessively light or dark.

If the exposure reading is made at some position above the microscope eyepiece, the image brightness is likely to be different from that at the film plane. This fact must be considered in calibrating an exposure meter. But again it does not really matter as long as light readings are taken at the same place and in the same way as in the test roll. Also, the light reading should always be made at the same distance above the eyepoint. A few manufacturers make adapters that provide for identical repositioning of the photocell.

Methods of Measurement. Wherever the reading is made, there are two methods of measuring brightness or exposure time. One method is to read just the background brightness with the specimen slide moved aside on the stage. This system provides a large area of uniform brightness on which to make a reading. The method is especially useful with reversal color film where exposure time is dependent on the brightest part of the specimen, which is essentially the same as the background brightness.

The second method of measuring brightness is to read the actual brightness of the specimen image. This system also works efficiently for color films as long as the specimen image has average brightness. If dense areas predominate the field, overexposure may occasionally be encountered, and the bright, highlight areas may appear washed out. This method should always be used, however, when negative films, either color or black-and-white, are to be exposed. This produces good rendition of detail in the darker areas of the subject as previously described in **Judging Exposures.**

Meter Calibration. Whichever method is used to make the light reading, the meter must be calibrated to conform to that method if accurate readings are to be made consistently. You must calibrate the meter to the method if some form of common light meter made for regular photography is adapted for microscope use, in which there are no *f*-stops. With the common light meter, you set the exposure index, observe the light level indicated (generally by a moving needle against a scale), and set the light level reading on a moving dial. This action results in a number of shutter speed/*f*-stop combinations that could be selected. But in the microscope there is no marked *f*-stop, so an effective *f*-stop for the system must be determined; that is one of the most important purposes of the test roll. With the initial test roll, you take a light-meter reading (at the side telescope, for example) and record it; at the moment it is meaningless. Then, you expose the roll using all shutter speeds, develop the film, and select the best exposure (from the negative once the skill has been acquired). Now go back to the meter computer. Set the speed of the film in use and then set the arrow or other indicator against the light value read. Find the shutter speed that represents the best exposure. Note and mark the *f*-stop value appearing opposite that point. This point is now the reference for all future uses of the computer, i.e., for determining the best shutter speed for a given light level.

Very simple and inexpensive light meters

can be made out of surplus solar batteries. Since they generate current when light strikes them, they need only be hooked up directly to an ammeter or microammeter.

Integrating Meters. Many current photomicroscopes designed for brightfield photomicrography, and others involving phase and interference systems, are equipped with photocells or photoresistors that integrate and measure the brightness of the image. This is integrated with other control data, like film speed, to time the exposure automatically. A wide range of brightness is accommodated. Allowances for filter factors are made by the integrating element. Film speed ratings, reciprocity departures, and subject density can be preset by the photographer. Those who practice photomicrography extensively would do well to study the various types of integrated systems offered by manufacturers. Automatic exposure determination devices were discussed in an earlier section on types of photomicrographic apparatus.

Exposure Record

When photomicrographs are made often, it is advisable to make a record of the conditions involved. A detailed record will permit duplication of a setup for future use either in rephotographing a particular specimen or in the photography of similar specimens. This information could be recorded on the envelope containing a photomicrograph or in a notebook.

The data may differ for some photomicrographs. Exposure data for black-and-white films, for example, should include the exposure index, specific developer, time, and temperature. If the eyepiece-to-film distance is always fixed, as in most eyepiece cameras, you may omit this entry. The purpose of the photomicrograph could also be included (for whom it was made and why it was made). Other data that might be included on exposure records include tube length (for microscopes with mechanically adjustable draw tubes), supplementary magnification setting, exposure meter reading, microscopical accessories, and photomicrographic camera.

Exposure Record for Photomicrography

NAME _____   FILM TYPE _____

DATE _____   EI _____

Exp. #	Specimen	Light Source	Filters	Cond.	Obj.	Total Mag.	Aperture Setting	Exp. Time
1								
2								
3								
4								
5								
6								
7								
8								
9								
10								
11								
12								
13								
14								
15								
16								
17								
18								
19								
20								
21								
22								
23								
24								
25								
26								
27								
28								
29								
30								
31								
32								
33								
34								
35								
36								

Chapter Eight
FAULTS IN PHOTOMICROGRAPHS

Even the most careful photomicrographer sometimes makes mistakes and produces a photomicrograph of less than the best quality. To correct such mistakes, you must recognize causes and effects. There are many factors that can affect image quality. Some are related to low-quality optics, while others are related to improper adjustment and alignment of the illumination and optics. The specimen itself can influence image quality; it may be too thick or be improperly stained. Dirt, dust, or grease on any of the optical components can affect the image. Incorrect use of filters in black-and-white photomicrography can reduce contrast or detail rendition. In color work, the wrong filters will affect color balance.

More-Common Faults

Unsharp Image

Lack of image sharpness is probably one of the most common undesirable effects in photomicrography. There are several causes.

Even when the image is focused sharply, there may be some slippage in the fine-focus adjustment on the microscope. If this happens often, the microscope may be too old or worn or the adjustment may be loose. You can find out whether the image focus changes by watching it for a short while after critical focus. If it stays in focus, something else is causing the trouble.

The camera shutter must not be actuated too hard or too fast, causing vibration that blurs the image. Always use a cable release and press it slowly. Mechanical isolation helps here.

If the image is focused on a ground-glass screen, it may be that the surface of the glass is too coarse, making critical focus difficult. Use a ground glass with a clear center spot and cross hairs so that focus is achieved on the aerial image. This situation was discussed under **The Ground Glass** on page 45.

Most eyepiece cameras include a reticle in the observation eyepiece. This reticle must be sharply focused. A young person's eyes will accommodate the focus of this reticle even when it is not in sharp focus.

Check before making the exposure to be sure that both the reticle and the specimen are in focus. If the reticle is out of focus, then the specimen image may also be unsharp when it is recorded on film. A focus adjustment for the reticle is provided on the observation eyepiece.

If a high, dry objective is used in the microscope, and if the cover glass on the specimen is too thin, undercorrection for spherical aberration is introduced. The image can never be focused sharply. There are only two remedies. (1) If the cover glass can be removed and replaced with one of correct thickness (No. 1½), a better image will be obtained. This remedy, however, is not always possible since the cover glass may be cemented firmly in place. (2) An oil-immersion objective of comparable magnifying power may be used to obtain a better image. Then the cover-glass thickness is less important. Objectives of this type are available from several firms. If one is to be purchased, make sure it will work efficiently with the microscope in use. Or use an objective with a built-in correction collar.

Avoid using objectives designed for a different microscope with a different tube length. This may introduce spherical aberration, affecting image sharpness.

People who wear glasses or contact lenses sometimes have difficulty achieving sharp focus in a microscope or through an observation eyepiece on a beam splitter. Check with the manufacturer who may have a means for solving the problem. Sometimes use of a high-eyepoint eyepiece can correct this problem.

Vibration will cause a recorded image to be unsharp, particularly at high magnification or with long exposure times. If any continuous bench vibration occurs, it will be noticeable in the microscope at high magnification; the image will appear to be in motion. A fast shutter speed (1/100 second or faster) or electronic flash will minimize the effect but will not remove the cause. If you can find the reason for the bench vibration, possibly you can eliminate it. The alternative is to mount the microscope or the stand on suitable shock absorbers. Suspect shutter vibration also. A slow shutter speed will minimize this effect because, if the cause arises in the shutter, the vibration is only prevalent during the first part of the exposure. A card in the light beam can be used for long exposures.

Hazy Image

An image appears hazy when a grease spot is present on the front lens of the objective, on the top lens of the eyepiece, or on the specimen slide. These surfaces may have been touched accidently. Inspect all glass surfaces frequently and clean them when necessary. Moisten a piece of lens tissue with 1,1,1-trichloroethane (EASTMAN Organic Chemical No. T3613) to clean a greasy surface. Avoid too much solvent, however, as this can affect the cement in the objective lens or the mounting medium on the slide.

A small drop of immersion oil on a dry objective will also produce a hazy, unsharp image. This is usually an accident and can occur when a dry objective is placed in position over a previously oiled slide. The oil can be removed with tissue moistened with 1,1,1-trichloroethane. Oil should always be removed from a slide after use.

Other causes of haziness or lack of sharpness include (1) uncoated objectives, (2) insufficient blackening in the objective or in the microscope body tube, (3) field diaphragm opened too far, (4) aperture diaphragm opened too far, or (5) too thick a specimen.

Uneven Illumination

When the objective and substage condenser are not aligned satisfactorily or when the illuminator and the light source within it are not correctly aligned with respect to the microscope, the illumination will be uneven. The effect will be quite noticeable in a recorded image. The background or the specimen image will be darker on one side than the other.

The objective in a microscope is usually fixed in position, so no adjustment of its position is possible. The substage condenser, however, can and should have centering screws so that it can be centered with respect to the objective. To find out if it is off-center, look down the tube of the microscope with the eyepiece removed. A circle of light will be seen that is the back focal plane of the objective. When the substage diaphragm opening is decreased, the circular image of the diaphragm appears. The image of the diaphragm should be centered. If it is not, adjust the screws on the condenser until it is.

When the objective and condenser are properly centered with respect to each other, but the illumination is still uneven, either the illuminator is off-center or the lamp in the illuminator is off-axis. Very often the lamp itself is off-center, particularly with built-in illumination. The position of the filament may vary in different tungsten lamps, even in those of the same type. Some microscopes with built-in illumination have centering screws on the lamphouse for centering the lamp. This centering can be checked by placing a piece of white paper in front of the field lens and adjusting the centering screws. When the lamp filament is off-center, the illumination on the paper will be uneven. Adjust the screws until the light is even. This effect may also be seen by looking in the microscope *without* a specimen in place; just the clear field will be visible. If the light is too intense, place a neutral density filter in the light beam to reduce the light level. The effect of uneven illumination, whatever the cause, is most noticeable at low magnification, since a large field of view is recorded.

Refer back to the section on setting up Köhler illumination.

Low Contrast

There are four principal reasons for low contrast in a photomicrograph. First, the substage diaphragm may be opened too far, creating flare and reducing image contrast considerably. The effect is noticeable in both color and black-and-white photomicrography. The setting of the substage diaphragm should be made according to the principles of Köhler illumination. (See page 28.)

Second, a filter may sometimes be used in black-and-white work for detail rendition with stained specimens, resulting in low contrast. If the color of the filter is similar to that of the specimen, it transmits the color of the specimen. The effect is to reduce visual contrast between the specimen and the background. Contrast in this case can be enhanced by (1) using a filter that has partial absorption for the specimen color, (2) using a film of higher than normal contrast, or (3) selecting a developer that produces higher contrast.

Third, the subject itself may exhibit very little contrast. This lack of contrast can be improved to some degree by the previously mentioned methods. Different lighting methods may improve the contrast also.

Fourth, the field diaphragm may be opened too far.

Less Common Faults

Too Much Contrast

Too much contrast is not likely to happen with color film, since processing conditions are fixed and contrast filters do not apply. It occurs in black-and-white photomicrography (1) when a high-contrast film is used and no effort is made to select a low-activity developer, (2) when a high-contrast developer is used with a regular film, (3) when development time is unduly prolonged, or (4) when a contrast filter is used erroneously.

Poor Resolution

Probably the principal cause of poor resolution is improper use of the substage diaphragm. When it is reduced to too small an opening, resolution is decreased considerably, artifacts and diffraction are introduced, and a generally poor image results. Follow the technique of Köhler illumination in adjusting the diaphragm to the correct opening.

If the substage condenser is not correctly adjusted and its position is too low, the effects are similar. The objective is not used at full numerical aperture.

Poor resolution also occurs when a photomicrograph is enlarged too much. Empty magnification results, and the image appears unsharp. When selecting the magnification, follow the rule of 1000 times the NA of the objective. At times, overmagnification may be permissible when the purpose is to make fine detail larger and therefore easier to see—a psychological benefit.

Bright Spot in Field

A bright spot in the field can occur when one uses a conventional camera with an integral lens or an eyepiece camera with a compensating lens above the microscope eyepiece. Ideally, the eyepoint of the ocular should be at, or very near, the front surface of the lens. If the front surface of the lens is too close to the ocular, an out-of-focus image of the back lens of the objective may be recorded as a bright spot when a reversal color film is used. It would, of course, be a *dark* spot on a negative, becoming a bright spot in the print. If the camera can be moved up a little farther above the microscope, the spot will disappear. If it is moved too far, however, image quality may suffer, or the image may be vignetted.

Shutter-Blade Image

If the camera shutter is too far above the eyepoint, a silhouette image of the shutter blades may be recorded as they open, particularly with fast shutter speeds. This phenomenon is known as *shutter shadow*. The shutter (leaf type) opens from the center outward, then closes toward the center. If the eyepoint of the ocular is positioned near the center, this effect cannot occur. Of course, if a lens is in the camera, one has little control of the eyepoint position, other than to place it at or near the front surface of the lens. In this case, fast shutter speeds (1/60 second or shorter) should not be used. If necessary, a neutral filter can be used to reduce the light, thus permitting slower shutter speeds. When the camera does not contain a lens, the eyepoint can be positioned in the center of the shutter. Occasionally, none of these methods works; in this case, changing to a high-eyepoint eyepiece always clears up the problem.

Out-of-Focus Spots

Out-of-focus spots are extremely common in photomicrographs. One of the most frequent causes of out-of-focus spots is dust on the cover glass of the specimen slide itself. The slide and cover glass should always be wiped clean just prior to photomicrography. The most common places for dust to settle and record as out-of-focus spots are the upper side of the field lens of the eyepiece and near the field diaphragm glass when the field diaphragm is mounted in the base of the microscope. Dusty filters near the field diaphragm are another source of out-of-focus spots.

Dust and dirt particles on the front surface of the lamp condenser lens or on the top surface of the eyepiece will also record as out-of-focus spots. They may be colored on a color film or gray (or black) on a black-and-white film. Scratches on a glass surface will record as unsharp spots. Bubbles in a lamp condenser or in a heat-absorbing filter in the illuminator will show as spots also; they may be bluish in a color film record. Pinkish or bluish spots may also be due to dust on, or bubbles within, built-in diffusion plates. Cracks in those elements will produce streaks.

Find the source of out-of-focus spots by rotating or otherwise moving all subjected optical components one at a time. Sometimes it is necessary to make photomicrographs while going through the rotating or moving procedure. Check all glass surfaces frequently.

Chapter Nine
ENHANCING SPECIMEN VISIBILITY

Optical Techniques

A majority of the subjects examined or photographed through a conventional brightfield microscope appear either dark or colored against a light background. When they appear dark, it is due to light absorption within their elements. When they appear colored, it is often because they have been treated with biological stains to produce color contrast. This color contrast is a result of differential absorption of various stains by the elements of the specimen.

In an unstained condition, many specimens exhibit little or no contrast when viewed in ordinary brightfield microscopy. They are colorless and comparatively transparent. Consequently, they are practically invisible. When staining such a specimen is either impossible or undesirable, conventional brightfield microscopy cannot be used. Another microscopical technique must be used to make the specimen visible. The particular method selected depends upon the specimen itself and the particular results desired. The following discussions present brief descriptions of those special techniques that will either produce better optical contrast between the elements of a specimen and its background or allow images to be viewed or photographed with finer resolution.

Both simple and sophisticated optical arrangements have been worked out to enhance the contrast between specimens and background and to delineate structures in transparent and translucent subjects. Optical and mechanical designs involve numerous slits, plates, prisms, and lens and condenser combinations. Photomicrographers can best manipulate these, and evaluate their results, when they have a grasp of the underlying principles. Manufacturers provide operating instructions, but interpreting the instructions requires some knowledge of the physics of light for complete understanding and application.

Fundamental knowledge provides another benefit. Many of the devices are quite intricate. Tolerances of the order of a wavelength of light are involved. Therefore, prudence demands that photomicrographers confine themselves to making only the necessary operating adjustments and alignments. Any other manipulations or repairs should be left to the manufacturer.

Darkfield Method

Many transparent and semitransparent specimens—such as microorganisms, cell structures, and crystal inclusions—are not readily visible in a brightfield microscope. Their visibility can be improved greatly by a method called *darkfield illumination*, in which the specimen is seen as a bright object against a dark, or even black, background.

For the darkfield method, the cone of light normally illuminating the specimen does not enter the microscope objective. In the darkfield microscope, only light that is scattered or reflected by the specimen enters the objective. This is achieved in the conventional microscope by use of darkfield diaphragm stops inserted underneath the condenser and with the aperture diaphragm opened to its maximum. The darkfield stop may be used at low and medium power with dry objectives.

Special darkfield paraboloid or cardioid substage condensers must be used at high power. The NA of the darkfield condenser must be greater than the NA of the objective. The cardioid and paraboloid condensers use the oil-immersion technique, which requires that oil be placed between the bottom of the glass slide and the top of the condenser. Great care must be taken that bubbles are not introduced either in the oil under the oil-immersion objective or in the oil between the slide and condenser. A single bubble will introduce considerable flare and reduce both image contrast and optical quality. Entrapped air bubbles can be seen by looking at the objective back focal plane.

Since special darkfield condensers are to be used with oil, particular care must be taken to obtain proper slide thickness. The correct range of slide thickness is usually specified on the condenser mount. If the slide is too thin, the oil layer collapses when the condenser is focused critically. If the slide is too thick, it is often impossible to obtain correct focus of the condenser unless a higher viscosity immersion oil is used.

The essential principle of the darkfield optics is the formation of a hollow cone of light whose apex occurs in the plane of the specimen. When the light is carefully focused

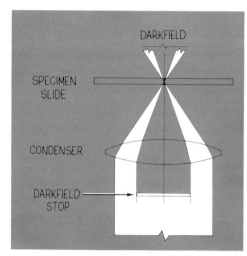

Fig. 9–1
DARKFIELD ILLUMINATION—*The central portion of the illuminating beam is blacked out to produce darkfield illumination. The opaque stop may be a separate element or part of a special darkfield condenser.*

at the plane of the object but no object is present, the hollow cone of light passed through the condenser produces no illumination in the microscope because the objective is inside the dark base of the hollow cone. When a specimen is present, the light is deviated, or scattered, into the objective by structures on the specimen. A bright image of these details is then visible against a dark background. Because of the high contrast of the image, the system is capable of detecting extremely small particles.

Color is seldom produced in a darkfield microscope, except in fluorescence work; thus, black-and-white films are widely used. Since considerable light is lost in this system, medium- to high-speed film, such as KODAK PLUS-X Pan Film or KODAK TRI-X Pan Film, is appropriate. An electronic flash may be necessary if the subject is in motion.

A darkfield microscope is an excellent tool for use in biology and medicine. It can be used effectively at high magnification to detect and photograph living bacteria. Similarly, at low magnification, whole mounts and tissue sections can be viewed and photographed. In marine biology, a darkfield microscope at very low power is used extensively for recording sea life such as algae and plankton.

Any brightfield microscope can be converted easily and quickly for darkfield work at low and medium power. One simply has to cut a circle out of opaque material, such as black paper or thin cardboard, of such diameter that the light is just prevented from entering the front lens of the objective when the stop is placed in the filter carrier beneath the condenser. The simplest way to determine the diameter of this opaque stop is to place a *transparent* metric rule in the filter carrier and look at the objective back focal plane by any of the methods described in the section on setting up Köhler illumination. At the objective back focal plane the rule will be seen, and the diameter can be measured directly by noting the number of divisions on the rule across the diameter of the back focal plane. Cut out a black paper circle of this diameter. Either mount it on a clear plastic, acetate, or glass disk and put it into the filter carrier or attach it to a glass slide and fasten the whole thing to the bottom of the condenser with pressure-sensitive tape. Alignment can be made by observing the objective back focal plane while moving the opaque stop about. Once the correct diameter has been determined and verified in use, a metal stop supported by a spider mount can be made for permanent, durable use.

Rheinberg Differential Color Illumination

Differential color illumination, particularly that form known as Rheinberg illumination, is a strikingly beautiful and important method of microscopical illumination. In principle, the method is an extension of darkfield illumination, the major difference being the substitution of transparent, colored central stops in place of the usual opaque stop, and the employment of transparent, colored, annular stops in the path of the usual white light. The combination of these colored central and annular stops results in specimens appearing in any color on a background field of any other color. Almost all objects, mounted and unmounted, whether stained or unstained, respond to differential color illumination. The method, sometimes called optical staining, is especially useful for unstained, transparent, and colorless specimens—living or nonliving— that ordinarily lack contrast because of the similarity of their refractive index to the refractive index of the mounting medium.

With Rheinberg illumination the central stop attenuates axial light and the transparent,

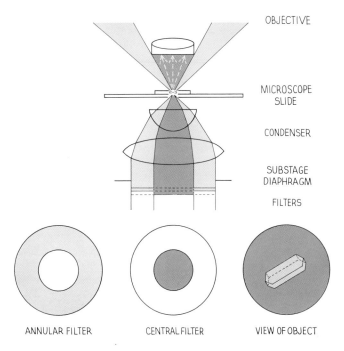

OBJECTIVE

MICROSCOPE SLIDE

CONDENSER

SUBSTAGE DIAPHRAGM

FILTERS

ANNULAR FILTER   CENTRAL FILTER   VIEW OF OBJECT

Fig. 9–2

RHEINBERG ILLUMINA-TION—*By introducing a colored central stop instead of an opaque one as with darkfield illumination, the microscopist can produce "coloredfield" illumination. A second color filter in the form of a peripheral annular stop provides color to the specimen so that it stands out against the colored field. The annular stop can be made up of segments of different colors to give multicolored effects. With low power objectives, the central stop may have to be reduced in size by an opaque ring stop.*

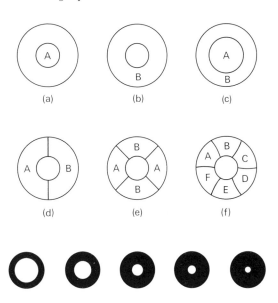

73

colored central stop will give its color to the background field of view. With only a colored central stop and the substage diaphragm opened, the specimen will appear white on a colored background. It remains only to add a ring or annulus of a different color to make the specimen colored as well. Figure 9-2 shows the placement of both central and annular colored stops, their position below the condenser, the light path through both filters, and the resulting appearance of the specimen in the color of the annular stop against a background of the color of the central stop.

Any combination of colors can be used, but in general it will be found desirable to keep the background a less intense color. In fact the background must also be reduced in light intensity level. Remember, the background color is coming through straight up, directly from the illuminator with no attenuation, whereas the light from the specimen is only that relatively small amount that is refracted, reflected, and diffracted. To increase the apparent light level of the specimen, the background light must be reduced. This can be done in a number of ways, including the addition of frosted or opalized glass, or WRATTEN Neutral Density Filters to the central stop color.

With both rings colored then, one may produce, for example, yellow objects on a red background, blue objects on a green background, yellow objects on a blue background, or any other combination that suits the immediate purpose. Furthermore, the annular stops may be multisectored so that a specimen can be illuminated with two, four, six, or eight different or alternating colors simultaneously from as many sides. Four-sector disks are invaluable in work with woven fibers or textiles since the warp and woof fibers will be differently colored. Figure 9-2 shows some of the possible arrangements of filter colors.

The same filters used for medium- and high-power work can often be used for low-power work as well if black, opaque ring stops are used to block out excess annular color. For example, if a particular set of filters is made to work for one objective, say 40X, and then a 10X objective is rotated into place, the central stop will be too large in diameter and will add its color to the specimen as well as to the background. If the eyepiece is removed and you look down the body tube, you will see only the central stop color. However, if you place a black ring over the central stop so as to cut out the central stop color extending beyond the objective back focal plane, you can use the stop with the lower power objective. Examples of some of these opaque rings are also shown in Figure 9-2.

Stop-Contrast Method

Darkfield illumination provides the means for the greatest enhancement of image contrast. It is especially valuable for particles but is extremely wasteful of light. Wilska worked out a stop-contrast method for improving the image contrast of unstained specimens. While his method was developed after the advent of phase-contrast systems (described in the next section), it is easier to understand and therefore is presented first. The basic scheme of manipulating direct and deviated beams is common to both systems.

The anoptric arrangement is quite similar to a brightfield system, but the image effect is one of darkfield (or semi-darkfield, as pointed out further on). A ring-shaped, or annular, slot is located near, or projected to, the condenser. This is illuminated by a solid-source lamp (ribbon filament or arc). The ring of light becomes the source for the microscope. It is focused, by the condenser and objective, near the periphery of the rear aperture plane of the objective (not in the plane of the primary image).

In the anoptric system, the hollow beam is partially blocked in the aperture plane by a *semiopaque* (usually 90-percent) ring coated on the rear element of the objective. The shadow of the ring mask matches the focused aperture-plane image of the ring source, thereby allowing only 10 percent of the illumination from the source to be integrated by the eyepiece.

A specimen can transmit some of the direct rays to the primary plane, where they are not completely masked off. Since a transparent specimen has such low structural opacity, the direct beam through it contributes only a little to the final image detail. As in a darkfield microscope, scattered light from the specimen also forms an image. In essence, a darkfield situation is presented to the eyepiece. The background brightness depends on the degree of opacity of the annular mask.

Contrast is improved partly because of this darkfield tone arrangement, but chiefly because central, direct illumination that would come up through a fully lighted condenser is not present with an annular source. Such central rays flood a field from all angles, whereas peripheral rays coming at an oblique angle provide modeling for the specimen. This modeling appears in the deviated beam, but would be too weak to be seen with transparent specimens were a central beam present.

In some anoptric arrangements the annulus is not blocked off. Instead, it is left clear, but the rest of the rear surface of the objective is coated—usually with a 50 percent opacity.

Thus the contrast roles of the direct and deviated beams are reversed. The first mode is useful for recording specimens of low visibility such as unstained chromosomes; the second, for highly refracting, bright subjects like yeast cells.

Varying the thickness (opacity) of the coatings that do the masking adjusts the intensity of the direct background beam and the amount of light deviated by the specimen. An even balance (equal intensities) provides the best delineation for most subjects.

Phase-Contrast Method

The phase-contrast microscope can be used to produce excellent contrast effects with a wide variety of otherwise transparent specimens. Since it permits visualization of interior details in cell structures, it has a definite advantage over the darkfield microscope. It is widely used in tissue culture study, where it permits one to examine and photograph living, growing cells. With time-lapse cine-photomicrography a specimen can be photographed at intervals timed to be synchronized with change. Projection at increased speed compresses the time. The study and photomicrography of living blood specimens by phase contrast also becomes possible.

The phase-contrast method was introduced by Zernike. Fundamentally, the procedure involves a direct and a deviated beam, as in the anoptric system. The difference is that the masking and balancing of the beams are accomplished by phase-interference modulation, rather than by opacity or amplitude modulation.

When a ray of light from a single point source is split in two and each of the two rays is passed through the same transparent medium, they can be recombined without interference. But if each separated beam passes through a medium of different refractive index, one will be speeded up or slowed down relative to the other. Then the two, when recombined, may be out of phase. If so, interference occurs and the recombined beam is not as intense as the original. To appreciate this, visualize the two split rays as waves traveling side by side. If neither one is altered, they can combine *in phase*. Peaks will meet peaks and valleys will meet valleys; there is no destruction of intensity. But if one component is altered in velocity, the waves may no longer match in configuration. When the configurations differ by less than a wavelength, the waves are out of phase and their recombination results in a loss of intensity. When they are out of phase by 1/2 wavelength, peaks will meet valleys and the rays will cancel each other out, extinguishing the beam.

In addition to refraction and retardation by different media, diffraction at edges and scattering from very fine details can also change the phase of the light waves. Diffraction is largely responsible for the edge effects seen in phase-contrast microscopy.

Again in the phase-contrast system, an annular source is utilized. But instead of an annular density mask behind the objective, a 1/4-wave ring is used. This ring retards a wave by 1/4 wavelength. As before, the specimen in the defocused direct beam deviates much of the light, depending on refractive index, thickness, and the fine structure. Again the deviated light goes to the eyepiece via the primary image. However, the deviated rays have been changed in phase by the specimen. The change plays a vital role in forming a high-contrast image, even though it may be as small as 1/20 wavelength.

When the beams of direct and deviated rays from the specimen are focused in the primary image, interference occurs between the two recombined, specimen-image components and between edges and background. The amount of intensity reduction in any given structure depends upon the phase difference of the two rays imaging that structure. When this is 1/2 wavelength, maximum reduction (darkness) occurs.

The brightness of the background (in a given setup) is always the same tone, corresponding to a medium gray in photography. But the brightness of the specimen image relative to this tone can be altered by several means. Phase plates and mounting medium control this aspect.

The image appearance, regardless of the optical elements used to modify it, is indicative of the contrast that has been obtained. When the specimen is generally lighter than the background, bright-phase contrast exists; when darker, dark-phase contrast exists. The outermost fringe, or halo, between the specimen and background is black in the first instance and white in the second.

Contrast can be manipulated in degree and phase to a practical extent by selection of the medium in which the specimen is mounted. If the refractive index of the medium is too close to that of the specimen, very little contrast will result. The medium should differ sufficiently in refractive index to provide adequate contrast for both visual examination and photomicrography. With tissue specimens, for example, a medium other than balsam is needed. Balsam is a common mounting medium, but its refractive index is very much like that of unstained tissue. A medium such as glycerin, for temporary mounts, or Diaphane and white corn syrup for permanent mounts, will provide much better contrast. The refractive index of balsam is about 1.53; while Diaphane and glycerin have a lower index, about 1.47. (See Common Mounting Media, Table 3, page 00.)

In phase-contrast photomicrography, a green filter is commonly used in conjunction with a black-and-white film. The green filter is appropriate because phase objectives are most often optically corrected for green light. An excellent filter for this purpose is KODAK WRATTEN Filter No. 58. A KODAK WRATTEN Filter No. 61 provides a slightly narrower band pass.

Color films are normally not needed since the phase-contrast method is most frequently employed with colorless, or almost colorless, images of low original contrast. Optimum phase-contrast effect is obtained with a single color, normally green. Color films may be used for photomicrography of very lightly stained objects or photomicrography with polarized light. If a slide for projection is needed, color reversal films are still the best choice for convenience and cost. An ortho, or green-sensitive, black-and-white film of moderate contrast is quite suitable for phase-contrast photomicrography. Of course, a pan-sensitive film will also work.

Interference-Contrast Methods

There are several other types of microscopes that utilize the interference effect. The most useful ones for general microscopy are those based on the interference-contrast principle. The method yields contrast enhancement just as a phase-contrast microscope, but it eliminates most peripheral and structural edge halos, thereby delineating fine detail. It also introduces color effects and gives a three-dimensional appearance of relief.

For interference-contrast methods, the condenser is fully illuminated, but apart from this there is a basic difference in the optics of phase- and interference-contrast systems. In the interference-contrast microscope, it is the optical system that produces the two interfering beams; it is not the specimen, as in the phase-contrast method. The interference-contrast beams are called the object beam and the reference beam. The first carries the image, and the second is either a homogeneous or an asymmetrical beam. Interference takes place between two types of beams, not between two image components, as in the phase-contrast system.

Once the object and reference beams are split, they are separate in the plane of the specimen. They are recombined for interference before they reach the eyepiece. There are

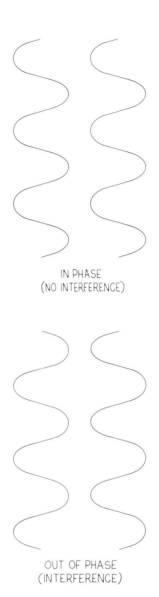

IN PHASE
(NO INTERFERENCE)

OUT OF PHASE
(INTERFERENCE)

Fig. 9–3
PHASE INTERFERENCE—
When light rays are combined in phase there is no interference. Out-of-phase beams cause destructive interference.

two commonly used methods for affording this separation. These are the *double-focus* system and the *shear* system.

In the first, both beams are focused by the condenser along the optical axis of the microscope. The object beam is focused in the plane of the specimen. The reference beam is focused above or below the slide and forms no structured image. In effect, the specimen is seen through the even blur of the reference beam. The two beams are displaced vertically. The wanted detail is imaged and interferes with the same detail out of focus.

In the shear system, both beams are focused in the plane of the specimen, but they are optically displaced laterally. In effect, only the subject beam illuminates the specimen—the reference beam is moved aside for subsequent recombination below the eyepiece. This beam should pass through a relatively clear area of the slide in order to minimize the imaging of unwanted detail, which would intrude on the subject detail. Thus this system is not as suitable as the double-focus method of studying tightly packed slides.

A commonly used interference system for all powers of objectives was worked out by Nomarski. It employs a modified shear system. The Nomarski system is very efficient in that it usually permits a full aperture. It is able to resolve finer structure than the phase-contrast method. However, the Nomarski image may give a deceptive appearance of surface relief where none exists. Depending upon whether the object beam or the asymmetrical reference beam dominates the image, structures appear to be above or below the surface. With a rotating stage and adjustment of the upper Wollaston prism, the microscopist can orient the slide to vary the effect.

Photography of interference-contrast images is similar to that used for brightfield illumination. It may be necessary, however, to close the aperture diaphragm further to obtain more even illumination. If this is done, the resolving power of the objective will decrease. Consequently, the best aperture setting is largely dependent on the specimen itself. Some experimentation will be necessary to obtain the best compromise.

Certain three-dimensional specimens in the metal, plastics, and glass industries and in crystallography and metallography can be examined by reflected interference with a system developed by Françon. A polarized-light microscope is adapted for axial illumination. Instead of Wollaston prisms, Savart plates, which are double flat plates of suitably cut quartz crystal, serve to produce divided beams. The degree of separation of the beams depends upon the thickness of the Savart plates chosen.

Polarizing Method

Polarized light has many valuable properties that merit understanding by the photomicrographer. Most illumination sources for microscopy emit beams of heterogeneous light; the waves are vibrating in all directions perpendicular to the axis of propagation. Polarizing media present optical slits that perform in a way somewhat like venetian blinds. They pass a portion of the light plane polarized according to the orientation of the slits. The other portion is absorbed. A polarizing element typically transmits about 32 percent of the incident light as plane-polarized light.

In the ordinary polarizing method, the microscope system is quite simple. A *polarizer* is placed below the specimen, usually in a rotating graduated carrier just beneath the condenser. An *analyzer* is placed above the specimen at the back of the objective, at the bottom of the drawtube, or within or even on top of the eyepiece. In true polarizing microscopes—also called petrological microscopes—the analyzer is in a rotating graduated mount in the body tube just a short distance above the objective nosepiece. There is no difference between the polarizing elements of the polarizer and analyzer; they are just given different names to distinguish them. They are ordinarily used in a crossed position, i.e., with their vibration directions perpendicular to one another so that the field of view which contains no specimen is black.

The underlying usefulness of the polarizing microscope lies in these facts. The vast majority of transparent materials will interact with the beam, causing the materials to appear lighted, and usually colored, against a black background. Some materials will have no effect on a polarized light beam, regardless of their orientation, and will remain dark between crossed polarizers.

Substances that do not interact and that remain black between crossed polarizers are termed *isotropic*. They turn out to have a single refractive index. Examples of isotropic substances are liquids (except liquid crystals), melts, unoriented polymers, unstrained glass, and minerals and chemicals in the cubic system of crystals, such as common table salt.

Substances that do interact and appear lighted and colored at some orientation between crossed polarizers are termed *anisotropic*. These substances turn out to have two or three principal refractive indices. Examples of anisotropic substances include all minerals and chemicals in the hexagonal, tetragonal, orthorhombic, monoclinic, and trigonal crystal systems, oriented polymers, cooled organic melts, and strained glass. Common table sugar is anisotropic. Biomedical specimens that exhibit anisotropism include hair, enamel, bone, fingernail, and muscle fibers in contraction.

One obvious benefit of the polarizing microscope, then, is to differentiate between isotropic and anisotropic transparent specimens, say as in a thin slice of rock for petrographic study. For the photomicrographer, this fact alone suggests one way of obtaining contrast with a colorless, transparent anisotropic specimen. Simply fully cross the polarizers and orient the specimen for maximum brightness.

But there is another, better-known effect produced with the polarizing method—interference colors. These colors are intrinsic to anisotropic material. They are determined by both the thickness of the material and the degree of anisotropism or birefringence. Birefringence is related to the numerical difference between the principal refractive indices. The birefringence of a substance cannot be changed without changing the substance, but the thickness generally can be changed; and thus the color of an anisotropic substance can be changed to another color by controlling its thickness. The colors themselves arise from retardation of one of the components of polarized light through the crystalline substances. Rays through the background of the specimen are extinguished, giving a black background.

The colors formed are not those of a continuous spectrum. Rather, they are interference colors formed in a definite sequence according to the path difference of the light beam through the birefringent material. (Similar colors are seen in soap bubbles or in a layer of oil on water.) The colors are in a characteristic sequence called Newton's series. Colors are formed in *orders* that, in the case of birefringent materials, must be interpreted according to material thickness and birefringence ($n_2 - n_1$).

Compensators, or tint plates, are thin slices of birefringent material—such as quartz, mica, and gypsum—that are cut in specially oriented crystallographic directions so that the directions of their fast and slow vibration components are known. These are inserted into slots in the body of the polarizing microscope so as to retard the light by a fixed or variable, but known, amount. The colors exhibited by anisotropic substances can be changed in two ways by the use of these compensators, depending on whether the crystal is in an additive or subtractive position with respect to the compensator. These compensators also aid in the detection and photography of substances with weak birefringence. The background color will be changed from black to gray or magenta-red, for instance, when compensators are introduced.

Any microscope can be used to obtain the effects of the polarizing method simply by placing inexpensive polarizing material below and above the specimen, and rotating one with respect to the other until a black field is obtained.

Very striking effects can often be obtained with a brightfield microscope equipped with polarizing elements. The formation and growth of birefringent chemical crystals is a fascinating subject that can be photographed on color film, with either a still or motion-picture camera. A low-power microscope is sufficient. Crystals are formed by dissolving a chemical in distilled water, placing a drop of the solution on a microscope slide, and spreading the solution thin. As the water evaporates, the crystal growth begins. Either of the polarizers can be rotated to produce the crossed position, which results in a dark background on which the colored crystal pattern forms. Sodium thiosulfate crystals work extremely well for this study. Organic chemicals are better prepared by melting them and allowing them to recrystallize. For further details on preparing crystal patterns for photography through the polarizing microscope, see page 23.

Other subjects that show polarization effects are rock sections ground thin for petrography, fibers, hairs, starches, and many components of plant and animal tissues.

Stereo Microscopy

When one looks at any object, the image recorded by the optical system in conjunction with the coordinating facility of the brain has a three-dimensional effect. This effect is the result of two different views of the object that are separated by an angle dependent upon the interpupillary distance of the individual but that are interpreted as one image. This is the principle of the Greenough binocular dissecting microscope, designed in 1897 for low-power work. In this type of microscope, each eye is provided with its own complete microscope. The paired objectives are mounted together at the approximate angle of binocular vision, 15 degrees. The rays from the objective pass through a Porro prism, which produces an image that is erect. The prisms in the paired objectives are similar to those used in field binouclars. The Greenough microscope has been modified and improved in recent years. The objectives have been

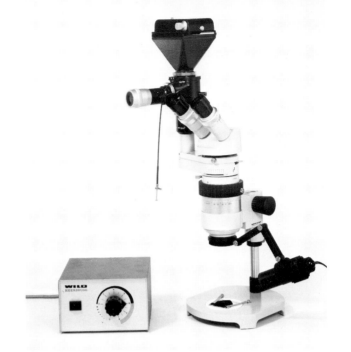

Fig. 9–4
STEREO MICROSCOPE –
Low power stereo microscopes with zoom optics are often equipped with attachment cameras for routine industrial use. Photo courtesy Wild Heerbrugg Instruments, Inc.

incorporated into a drum, or sliding nose-piece. Wide-field oculars and prisms produce a large, flat field of view with magnification ranging from 3X to 100X. Such a microscope has the advantages of great depth of field, long working distance, large and flat fields, and simplicity of operation.

The *stereo microscope,* as it is commonly called, has become invaluable in industry, medicine, teaching, or wherever a microscope of low power is needed.

Most newer stereo microscopes now have zoom capabilities and a maximum magnification of over 200X. The use of stereo microscopes varies from general scientific investigation in all fields to production examination in quality control. The most important use in relation to photomicrography is in examining the specimen to determine the best area to photograph. This is especially necessary in photomicrography of biological specimens—where dissecting, staining, and selective mounting are performed.

The stereo microscope has several limiting factors when specimens are to be photographed through one of the body tubes. The principal limiting factor to producing quality photomicrographs with stereo microscopes is the low NA of most objectives. Additionally, since only one objective, at a slight angle to the specimen, is used, the depth and resolution typical of stereo views is not recorded on

film. Many microscope manufacturers have accessories to correct this situation. Paired stereomicrographs can be made in several ways that are described in texts such as *The Practical use of the Microscope* by G. H. Needham.

The eyepieces of low-power stereo microscopes are generally of larger diameter (about 30 mm) than those of the compound microscope. The manufacturers of most photomicrographic attachment cameras make adapters to fit their cameras onto the wider eyepiece tube of the stereo microscope. A trinocular stereo microscope is convenient but not necessary for low-power photomicrography. The camera attachment can simply be placed in one of the two vertical or inclined binocular tubes. The attachment cameras are focused in the same way as described in an earlier section for high-power microscopes.

In addition to single-frame, low-power photomicrographs, one can make *stereo pairs* by photographing through one eyepiece on one frame, advancing the film, and photographing through the other eyepiece on the next frame. The resulting photomicrographs can be mounted and viewed in a stereo viewer where the specimen will be seen as three-dimensional. Correct orientation of the pairs is important.

Stereophotomicrography is also possible in several ways with the single objective of a monocular compound microscope. The most convenient way to make stereo pairs, and still maintain full numerical aperture, is to use a commercially available *tilting stage.* This stage tilts the specimen slide about 7 degrees to one side for one photomicrograph, then tilts it about 7 degrees in the other direction for the second photomicrograph. The resulting stereo pair is mounted in a single stereo mount for viewing in a stereo viewer.

Photographic Techniques with Special Illuminants

Fluorescence Photomicrography

Some materials will emit light of a longer wavelength when excited by short wavelengths of radiation. This phenomenon is called *fluorescence.* Ultraviolet radiation and blue light are often used as exciting radiation to produce visible light of longer wavelengths. If a substance does fluoresce, the effect is called either *primary fluorescence* or *autofluorescence.* Some materials do not fluoresce by themselves but can be impregnated with chemicals, such as certain dyes that will fluoresce. Dyes of this type are called *fluorochromes;* the effect in the original material is called *secondary fluorescence.* For example, when excited with ultraviolet radiation, chlorophyll will fluoresce with a deep red color. When stained with dilute acridine orange, human epithelial cells will glow orange-red under ultraviolet radiation.

Fluorescence microscopy and photomicrography are important in cytology. An application is in early detection of cancer in smears, exudates, and tissue sections. Fluorescence techniques are used in other fields of medical and biological research also.

A very efficient light source for most fluorescence work is the high-pressure mercury-vapor lamp which emits very bright radiation in both ultraviolet and short-blue wavelengths. When a light source having a continuous visible spectrum as well as ultraviolet radiation is needed, the xenon arc serves the purpose very well. Most fluorochromes are excited by ultraviolet radiation and produce fluorescence somewhere in the visible spectrum. Maximum fluorescence from dyes commonly used in antibody techniques comes

as a result of absorption of long blue wavelengths, however, and require a continuous source, such as a xenon arc or high wattage tungsten-halogen lamp.

When a substage mirror is used, one with an aluminized surface should be selected. Silver, often coated on these mirrors, is a poor reflector of ultraviolet radiation.

When ultraviolet radiation is used to produce fluorescence, such as with a mercury arc, an exciter filter is needed in the light beam to transmit this radiation freely and to absorb visible light not needed in producing fluorescence. KODAK WRATTEN Filter No. 18A (glass) has a high transmittance for the 365 nm (ultraviolet radiation) line of the mercury-vapor spectrum. It absorbs all visible light and appears black to the eye. A barrier filter is also necessary. This type of filter should absorb the ultraviolet radiation that is transmitted by the specimen and transmit the visible fluorescence. The selection of an efficient barrier filter can be critical. A KODAK WRATTEN Filter No. 2A or 2E can be used for this purpose. Although each filter will completely absorb ultraviolet radiation, the difference is in their absorption of short blue wavelengths. If in doubt, try the No. 2A Filter.

In a fluorescent-antibody technique, where fluorescein dyes such as fluorescein isothyocyanate (FITC) are used as fluorochromes, secondary fluorescence usually occurs in the green at about 540 nm. Maximum absorption of these dyes is at 480 to 490 nm in the blue region of the spectrum. A filter that transmits blue freely is normally employed as an exciter filter. KODAK WRATTEN Filter No. 47B is often used. A barrier filter for this application must transmit wavelengths longer than blue and must absorb blue completely. A KODAK WRATTEN Filter No. 12 or No. 15 will suit this purpose; both filters are yellow and will transmit the green fluorescence color freely.

When fluorochromes are used to stain specimens, a certain amount of short-blue autofluorescence occasionally occurs also, particularly with tissue selections. This autofluorescence often must be absorbed to avoid degrading the color of the secondary fluorescence produced by the fluorochrome. Both the No. 2A and No. 2E Filters absorb short-wavelength blue radiation, the No. 2E having the greater absorption.

A barrier filter must be used behind the objective. If it is not, the residual ultraviolet or short-wave length blue radiation will record as blue on color film and will degrade all fluorescence colors. A barrier filter of appropriate size can often be used directly behind the objective in the microscope body

tube, or a smaller-size filter can be placed in or on the eyepiece of the microscope.

The selection of a mounting medium is important. A medium must be chosen that has little or no autofluorescence. Autofluorescence of a mounting medium may be either pale blue or pale green in color and will degrade fluorescence colors. Temporary mounts can be made with either pure glycerin or Cargille's immersion oil, type A, or crown oil. These have very low fluorescence. If glycerin containing such impurities as acrolein is used, a light greenish autofluorescence may occur. Most permanent mounting media also produce autofluorescence; therefore, they should not be used in fluorescence work.

The pale-blue autofluorescence of a mounting medium can be absorbed to some extent by either a barrier filter, that absorbs some blue, or a pale yellow KODAK Color Compensating Filter, such as a CC20Y. Pale green autofluorescence can be neutralized by a pale magenta color compensating filter, such as a KODAK Color Compensating Filter CC20M. Heavier filters may also absorb some of the desirable fluorescence color. Quite often the KODAK WRATTEN Filter No. 2E will serve to absorb the ultraviolet and the pale-blue autofluorescence of a mounting medium as well.

Darkfield illumination is useful for both fluorescence microscopy and photomicrography. By this technique, the darkest background is achieved so that fluorescence colors stand out brightly. As previously stated, oil-immersion darkfield condensers are most efficient for medium and high power. Take great care that no bubbles are introduced into the oil and that the microscope slide is of the correct thickness for the condenser in use. A darkened room is essential to efficient darkfield work in fluorescence microscopy in order to exclude extraneous light.

For successful fluorescence photomicrography

- Use an objective with high NA. Select a low power eyepiece. Limit total magnification.
- Seek contrast rather than intensity.
- Illuminate the specimen from the top rather than by transmitted light.
- Use high-wattage tungsten-halogen lamps for routine work.
- Choose exciter and barrier filters to suit the fluorochrome.
- Remove beam splitters and prisms when making exposures.
- Work quickly to avoid quenching of the initial fluorescence.
- Use a high-speed daylight-type color film.

- Push process films only when higher film speeds are essential.
- Make exposure tests. Compare empirical results with indicated exposure of light metering devices. Center fluorescing cells in spot of spot metering device.

Even though a fluorescent image may appear bright to the eye, long exposure times are often necessary in photomicrography. Use color films to record the image approximately as it appears to the eye. A high-speed film—such as KODAK EKTACHROME 400 Film (Daylight)—will minimize exposure time and will record most fluorescence colors with reasonable accuracy. Use daylight-type film because of its balanced sensitivity to red, green, and blue. This film can also be specially processed for higher speed should it be necessary. Other daylight-type, color-reversal roll films of lower speed can also be used to record particular fluorescence colors.

Now widely used for fluorescence photomicrography is *vertical* or *epi illumination*. Efficiency of irradiation is improved since fluorescence is stimulated on the surface being viewed.

Ultraviolet Photomicrography

Since the limit of resolution attainable in photomicrography depends upon the wavelength of radiation, the highest resolution is obtainable with ultraviolet radiation. It is possible to make photomicrographs in the near ultraviolet (365 nm) by using a conventional microscope with a glass lens. Apochromatic objectives perform best. Optical glass, however, transmits little ultraviolet radiation below a wavelength of about 330 nm. To transmit the shorter wavelengths, a microscope must be equipped with fluorite optics or reflecting optics. A principal advantage of this type of microscope is that it can be focused on an image in visible light and maintain the focus throughout the ultraviolet region. A light source such as a low-pressure, mercury-vapor lamp is needed to provide radiation in the ultraviolet region. For photomicrography, a monochromator can isolate a particular wavelength or narrow band of wavelengths in the ultraviolet region. A KODAK WRATTEN Filter No. 18A can be used to isolate the 365 nm wavelength.

Quartz optics, other types of microscopes, and alternate illumination sources as well as a means for focusing the invisible ultraviolet radiation are discussed fully by Loveland (1970).

Photomicrographs in the near ultraviolet region can also be made with this ultraviolet microscope. It provides a higher degree of resolution than is obtainable with a conventional microscope. In addition, the ultraviolet microscope allows one to study the selective absorption of ultraviolet by living cells, tissues, and fine particles.

Initially, all photographic emulsions have an inherent sensitivity to blue and ultraviolet radiation. Sensitivity in the ultraviolet actually extends far into this region, but the response of a film or plate is somewhat limited due to absorption of ultraviolet radiation by gelatin. Many current color films have a special UV-absorbing overcoat on top of the emulsion which confines their sensitivity to the visible region. For long-wave ultraviolet photomicrography, almost any film or plate without a UV overcoat can be used to record the image. However, only black-and-white films need to be considered, since color films have no advantage.

Images formed by ultraviolet photomicrography are low contrast. Therefore, if a high-contrast film is not used, medium- to high-contrast development of conventional films may be needed. To obtain medium to high contrast, either extend the recommended development time or use developers that produce higher contrast. Note that blue- and blue-green-sensitive films are generally designed to provide higher contrast than panchromatic materials.

When higher sensitivity to ultraviolet radiation is needed, or when sensitivity is desirable in medium-wave and short-wave regions of the ultraviolet, it may be necessary to resort to special emulsions. Ask your dealer in photographic equipment and supplies about special materials for photomicrography and ultraviolet recording.

Infrared Photomicrography

Most subjects are examined through a microscope by transmitted light. Some subjects, however, are relatively opaque when placed under a microscope; few, if any, details are visible. Increased transparency can often be obtained by the use of infrared radiation, the long wavelengths beyond the visible spectral range. Infrared photomicrography is especially useful in entomology. Many insects have dark-pigmented structures that are opaque to visible light but are penetrated freely by infrared radiation. Heavily stained specimens, many textiles, forged and altered documents, dark-colored crystals, silicon wafers, and many other subjects can be photographed with infrared radiation. Infrared color photography has been used for differentiating biologic pigments, tissue structures, and inclusions, as well as for criminal detection methods.

Because of the long wavelength of infrared radiation and because of lens aberrations in that region, the infrared image cannot be as sharp as in ordinary photomicrographs. The benefit comes from the penetration of many visually opaque specimens.

Since normal exposure meters have low sensitivity in the infrared, exposure times are usually determined by trial. It is common practice to make a series of exposures, either on sheet film (by withdrawing the dark slide by definite amounts and varying exposure times) or on 35 mm roll film (by making several exposures on successive frames).

Before making infrared photomicrographs, test film holders and plateholders, including the draw slides, to make certain that they are opaque to infrared radiation. Failure to do this may result in fogged emulsion and complete deterioration of image contrast. Processing recommendations for infrared films are included with the materials. Even infrared films in 35 mm magazines should be loaded into the camera in total darkness since the velvet-lined lip is not IR-tight.

Because infrared photography is quite specialized, it is not feasible to go into details here. The basic requirements have been outlined and there is no particular complexity involved. Nevertheless, certain careful but straight-forward procedures are needed in focusing the photomicroscope. When infrared color film is employed, special attention must be paid to light sources and color balancing with filters. Those who wish to carry on the technique are referred to Kodak Publication M-28, *Applied Infrared Photography*.

Chapter Ten

PHOTOMICROGRAPHY OF OPAQUE SPECIMENS

Today, the methods and principles of the metallographic microscope and metallographic photomicrography have been extended far beyond their original purpose. The modern industrial laboratory has applied the methods and techniques of metallographic microscopy to the study of opaque, nonmetallic specimens as well. These include microelectronic circuit boards and their components, metal shadowed replicas, ores, ceramics, painted and otherwise coated or corroded surfaces, radioactive components, synthetic fibers and animal hairs, cosmetics, plastics, industrial dusts, combustion products, coal, and a host of new products of materials science.

Additionally, the general methods of using reflected light (often referred to as epi or incident illumination) formerly thought of only in terms of metallographic application are also extensively used in the biomedical field, particularly when working with tissue cultures, microbiological cultures, and wood specimens. Because of these wide applications of the metallographic type of microscope to fields outside of metallography, the photomicrographer should become familiar with the techniques of sample preparation and illumination used in metallography.

Metallography

Metallography is the study, interpretation, and recording of the physical structure of metals with the microscope. Metallographic study of the structure of a metal includes grain size, constituents, and foreign particles within the metal. Photomicrography is widely used in metallography for recording microstructure.

Metal Specimen Preparation

The preparation of metals for photomicrography usually requires mounting, polishing, and etching. These techniques differ when the metals and alloys are ferrous or nonferrous, or their texture ranges from extremely hard to very soft. As a result, the preparation of each metal specimen involves a slight modification of basic technique.

Occasionally it may be possible—indeed, necessary—to look at a metal specimen directly without any preparation whatever.

Spring clamps are frequently used to hold such specimens. Ferrous specimens can be positioned with small magnets. Most specimens, however, require full treatment.

Small or awkwardly shaped specimens can be handled more easily during preparation if they are embedded in a suitable material. The simplest technique for mounting specimens is a clamp made of material similar to the specimen, and designed so that the clamp jaws and specimen surface can be prepared together. Other methods include the use of thermoplastic or thermosetting synthetic resins in a metallurgical mounting press.

After embedding, the specimen surface is ground or filed, and then polished to reach the desired mirror finish. This finish must be free from surface deformation, and it is essential that the techniques used remove material by a cutting, rather than a buffing, action.

When the required mirror finish has been obtained, the microstructure of the specimen is revealed by swabbing the specimen with, or immersing it in, etching reagent. In general, etchants either attack the grain boundaries between constituents or stain the various constituents present. The exact choice of etchant depends upon its chemical relationship with the metal or alloy.

More information on methods of preparation and etchants to use on metal samples can be found in publications from the American Society of Metals, in standards of the American Society of Testing and Materials, or in books from the reference list.

Reflected Light Microscopes

Any type of microscope can be used to photograph metals or opaque objects. A biological microscope can be used at low magnification with oblique lighting. However, an integral light source, an adjustable stage, and the proper accessories are found in the vertical-type (Figure 10-1) and inverted-type metallographic microscope (10-2).

The Le Chatelier type of inverted microscope is commonly used in metallography. With this microscope, the specimen is held perpendicular to the optical axis of the objective by placing it face down on the stage. The

stage can then be moved up or down by both coarse and fine focusing adjustments. The objective is held in a vertical position beneath the stage. Alignment of the camera, microscope, and illuminator is comparatively permanent for speed and accuracy in production work. Various light sources are adaptable for use with this type of microscope.

For rigidity and also for convenience, the camera, inverted microscope, and light source are sometimes mounted on a horizontal stand or are a complete unit. Most types of metallographic microscopes include built-in facilities for polarized-light and darkfield work, as well as having the conventional brightfield arrangement. Accessories are available for phase contrast and for interferometry. Inverted metallographic microscopes are made commercially by several manufacturers.

Objectives

The objectives commonly used in photomicrography of metals are essentially the same as those used in regular transmitted-light photomicrography, the principal differences being their correction for use with objects *without* cover glasses and their use with microscopes having specific tube lengths. Cover glasses are not used with metallurgical specimens under vertical illumination because a considerable amount of light would be reflected at the glass surface and would result in poor illumination of the metal surface and obliteration of detail in the specimen in the photomicrograph.

You will recall that objectives for transmitted light have spaces between the front and rear lens element that are very carefully selected to rid the system of spherical aberration when used with a very specific cover-glass thickness. In making objectives for metallurgical use, the same lenses are used, but the spacer is selected for an infinitely thin coverglass, i.e., no cover glass at all. If these objectives are used for covered specimens, poor image quality will result. Objectives for use without cover glasses will usually be engraved so (or abbreviated NCG for no cover glass); German-made objectives may sometimes simply have the letters o.d., standing for *ohne decke* (without cover). On objectives where it is common to indicate the cover glass thickness, e.g., 40/0.85/0.17 (40X

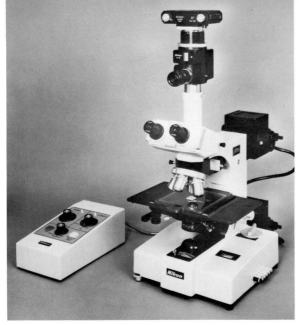

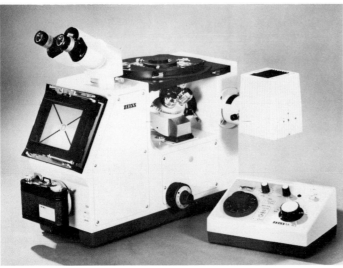

Fig. 10–1
METALLURGICAL MICRO-SCOPE—*A typical microscope provides for incident reflected light illumination as well as transmitted light when this is needed.* Photo courtesy of Nikon, Inc.

Fig. 10–2
INVERTED MICROSCOPE—*An integrated camera microscope for metallography has objectives under this stage. This microscope provides for either 4 x 5-inch or 35 mm film exposures.* Photo courtesy of Carl Zeiss, Inc.

magnification; 0.85 numerical aperture; 0.17 thick cover glass), a dash usually appears, e.g., 40/0.85/−.

Metallographic objectives are often corrected optically for a tube length of 185 to 250 millimetres to account for the addition of the vertical illuminator. (The standard for most biological microscopes is 160 milli-metres.) Commercial models of metallograph-ic microscopes have specified tube length built in, so you can ignore this factor as far as the microscope is concerned. Be certain, however, that the objectives are corrected for use with the inherent tube length of the microscope. The tube length for which an objective is designed is usually imprinted on the objective mount. If it is not, write to the manufacturer and verify the optical properties of the objective. If an objective is purchased specifically for metallography, chances are good that its tube-length designation is correct. It is not advisable to use an objective specified for a given tube length on a micro-scope designed with a different tube length because very poor image quality can result.

For this reason, objectives of one manufac-turer may not function efficiently on a micro-scope of a different manufacturer.

Some microscope objectives, particularly those used on the longer, horizontal metallo-graphs, are infinitely corrected. Rather than being restricted to a fixed, mechanical tube length, they can be used in mechanical config-urations that greatly exceed the usual 160 mm. These objectives are usually desig-nated so by an engraved infinity symbol (∞) somewhere in the objective legend.

If metallographic objectives—i.e., those corrected for use without a cover glass and for a longer-than-normal tube length—are used for ordinary transmitted light microscopy of covered specimens at a 160 mm mechanical tube length, poor image quality will result. For this reason, metallographic objectives are usually given a different thread or mounting device so that they cannot be fitted to the nosepiece of a transmitted-light microscope with RMS* thread.

*Royal Microscopical Society.

All comments regarding type (achromat, fluorite or semi-apochromat, and apochromat) and correction of objectives, including chro-matic and spherical aberration and field curvature, in the earlier discussion of trans-mitted light objectives apply to metallo-graphic objectives. See pages 5-10.

Eyepieces

The eyepieces used in metallographic micro-scopy are the same type, magnification, and design as those described earlier for trans-mitted-light use. See page 11.

The Aperture Diaphragm

Unlike the aperture (substage) diaphragm in the transmitted-light microscope, the aperture diaphragm in the metallographic microscope is usually an integral part of the vertical illu-minator and is used to control the cone of light entering the objective lens. In use, it should be so adjusted that the fullest aperture of the objective is utilized consistent with good, glare-free images. This adjustment is accomplished by looking into the microscope tube with the eyepiece removed. (An image of the specimen must be in focus in the micro-scope.) An image of the light source and of the aperture diaphragm is visible at the back focal plane of the objective. When the diaphragm is adjusted correctly, its image should fill the back lens with light. The edge of the diaphragm is then just visible within the periphery of the back focal plane.

The physical location of the aperture diaphragm is nearer to the lamp and farther from the microscope than the field diaphragm. (See Figure 10-3.) Note that this is opposite to the transmitted light microscope in which the aperture diaphragm in the substage condenser assembly is closer to the microscope than the field diaphragm that is associated with the illuminating apparatus.

The Field Diaphragm

The principal function of the field diaphragm is to minimize internal glare and multiple light reflections within the microscope by delimiting the illumination to the actual field under observation, and no more. The image will then have the best attainable contrast. The field diaphragm is located closer to the microscope than the aperture diaphragm (Figure 10-3). The objective itself acts as a condenser to image the field diaphragm in the plane of the specimen when the specimen is in focus. When you look either in the micro-scope or at the camera ground glass, you see an image of the field diaphragm simulta-neously with the microscope image.

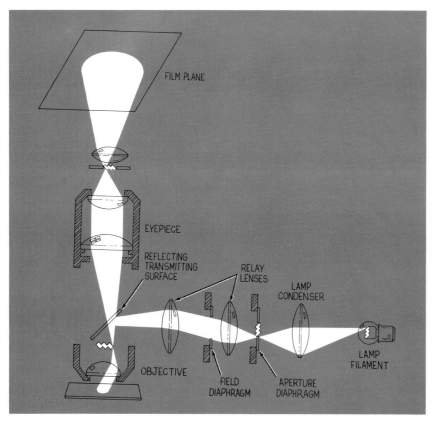

Fig. 10–3
INCIDENT ILLUMINA-TION—*Setting up Köhler illumination with reflected incident light requires the same manipulation of aperture and field diaphragm as for transmitted light. However, the arrangement of these elements differs from the transmitted light system. Note that it is the* field *diaphragm that is nearer to the objective while the aperture diaphragm is closer to the lamp condenser.*

Cameras

Several kinds of cameras are used in photomicrography of metals. They include:

- The horizontal or vertical open-back view camera with expandable bellows. This is aligned with the optical system.
- Integral cameras in commercial metallographic stands. Some use glass plates or sheet film; others use roll film.
- 35 mm single-lens-reflex or roll-film cameras adapted to the microscope. More information is readily available from microscope manufacturers.

Magnification

Image size in metallography is determined by the magnifying powers of the objective and eyepiece, by the microscope tube length, and by the bellows extension (eyepiece-to-film distance). An objective with a magnifying power of 20X will produce a visual image 200 times as large as the specimen when the eyepiece has a magnifying power of 10X. If this image at 200X is projected through the eyepiece to a ground glass or a film and focused at a distance of about 250 millimetres (10 inches) from the exit pupil of the eyepiece, the projected image will also be at 200X. Greater magnification can be obtained by drawing the ground glass farther back.

Magnification then will be directly proportional to the distance from the ground glass, or image plane, to the eyepiece. The distance is called the *bellows extension*. At 500 millimetres (20 inches) from the eyepiece, and with the above combination of eyepiece and objective, the image size would be 400X.

It is always necessary to determine magnification in metallography. This can be done with a *stage micrometer*, which consists of a microscope slide containing finely ruled lines with a finite separation in decimal parts of either inches or millimetres. In use, it is necessary only to place this slide on the microscope stage and to project an image of the lines to the ground glass of the camera. By comparing the separation of the lines on the ground glass with the original separation, you can measure image magnification directly. This image of the micrometer rulings can also be recorded photographically for more precise measurement later.

The American Society for Testing and Materials recommends that all photomicrographs of metals and alloys be made and reproduced in papers and journals at one of the following standard magnifications: 1X, 5X, 25X, 50X, 75X, 100X, 150X, 200X, 250X, 500X, 750X, 1000X, 1500X, and 2000X.

Illumination Systems

Incident Brightfield Illumination

A specimen to be examined or photographed in metallography is usually completely opaque and must be illuminated by incident light. This illumination can be directed at the specimen from any angle to produce a variety of effects, but most often vertical illumination is used. In this system, an illuminating beam is directed as nearly as possible along the optical axis of the microscope's objective lens to the specimen. For medium- and high-magnification metallography, the light source is usually at one side of the microscope and projects a horizontal focused beam to the optical axis, where it is reflected to the specimen. The reflector can be an integral part of a commercially made vertical illuminator and can direct light through the objective lens to the specimen. In low-power photomicrography, where the objective lens has sufficient working distance, a separate reflector can be placed between the objective lens and the specimen.

Either a clear, plane-glass reflector or a prism can be used in the illuminator. Some vertical illuminators are equipped with both a reflector and a prism, which are interchangeable. A prism gives far brighter images with more contrast than a plane-glass reflector, but unfortunately, a prism also reduces the aperture of the optical system. For medium- and high-power metallography, the plane-glass reflector is much preferred because of improved image quality. When the plane glass is used, however, only a small percentage of the illumination is reflected to the specimen, so an intense light source is required.

Vertical illuminators can be purchased as accessory units and fitted to an ordinary biological microscope. Such illuminators often have an integral light source and are usually suitable only for visual work since light intensity is very low. It is often necessary in photomicrography to use another light source of high intensity with the illuminator. The vertical illumination is also referred to as *incident brightfield illumination.*

Incident Darkfield Illumination

With increasing magnification, it is impossible to illuminate a specimen sample directly with a light source because of the short working distance of the objective. However, an even, overall illumination can be obtained by use of a special objective with an annular condenser built around it. Such equipment is available from several microscope manufacturing companies. The path of the light beam is down the outer periphery of the objective, through the annular condenser to the sample, and then back up the objective to form the image. This method is called *incident darkfield illumination.*

Lighting Surface Topography

If study of surface topography is necessary, as in metallography, it is enhanced with specularly reflected light, oblique illumination, or special light-interference techniques. When the light source is separated from the vertical illuminator, any pronounced movement of the objective will destroy the illumination adjustment in the microscope. It is therefore a practical necessity to focus an image by moving the specimen. You will need a vertically adjustable stage to provide specimen movement.

Some special techniques can also be adopted for opaque specimens. The simplest of these is *oblique* illumination. This can be obtained by decentering the aperture diaphragm. It is a good method for examining rough or scratched surfaces to show depth. Placing an opaque disk just in front of the

aperture diaphragm will produce *conical* illumination. Other techniques requiring specially equipped or adapted microscopes are darkfield illumination, phase-contrast illumination, the use of polarized light, and fluorescence photomicrography.

Light Sources

Due to the inevitable light losses in a vertical illuminator, the source of illumination should have relatively high intensity to render a projected image of reasonable brightness and clarity. The *carbon arc* has been a common light source in metallography. This arc lamp, consisting of a vertical (negative) and a horizontal (positive) carbon electrode, can be operated on either alternating or direct current. Direct current is preferable because it provides a steadier light source of uniform intensity; alternating current causes an annoying flicker that is objectionable both for visual examination and for photomicrography. Xenon arcs are provided on many modern microscopes and frequently fitted to older ones.

The *xenon arc* is an excellent light source for metallography. The arc is produced across tungsten electrodes in a glass envelope that contains highly pressurized xenon gas. The emission of the xenon arc is continuous not only in the visible spectrum but also in the long-wave ultraviolet region and in the infrared spectral region. This lamp produces illumination of high intensity and daylight quality (color temperature is 6000 K), a feature that allows the use of daylight-type color films with little or no filtering.

Tungsten-filament lamps can also be used in metallography. Although they are considerably less bright than carbon arcs, they are much steadier and do not require frequent replacement as do the arcs. A ribbon-filament lamp that is rated at 6 volts and 18 amperes and uses alternating current through a step-down transformer is quite suitable for everyday black-and-white metallography. When images of very low brightness are obtained, such as at high magnification or in polarized-light work, extremely long exposure times are encountered and a much brighter light source is appropriate. A suitable illumination device is the modern *tungsten-halogen lamp.* The tungsten-halogen lamp is available in 12-volt, 100-watt size. It emits efficient high-intensity illumination because the coil filament is small and compact. Its bulb life is several hundred hours, and it can be replaced easily and inexpensively.

The *zirconium concentrated-arc lamp* is another excellent light source for this work. Although its brightness is not quite as high as

that of the carbon arc, it is completely steady and will operate efficiently without replacement for a considerable number of hours. Also, its color temperature is 3200 K.

Köhler Illumination With Incident Light

The principles of Köhler illumination, which are commonly applied in standard transmitted-light photomicrography, can also be applied to photomicrography of metals. All metallographic microscopes contain two variable diaphragms—an aperture diaphragm and a field diaphragm, as described earlier. Both are used in adjusting illumination to enhance the quality of the image in the microscope and in the camera. The locations of these diaphragms will differ somewhat in the different models of metallographic stands, but will generally appear in relation to one another as they do in Figure 10-3. Setting up Köhler illumination for incident light (vertical illumination) is different in some respects from the procedure for transmitted light, so we will go through the steps one at a time.

Before attempting to set up Köhler illumination with incident light, reread the steps for setting up Köhler illumination by transmitted light, pages 28-35, because the same principles are applied to incident light.

Starting with a 10X objective and the plane-glass reflector in place, select a decidedly specular specimen, such as a highly polished piece of metal or star test plate. The initial setup of Köhler illumination will be much easier if such a specimen is chosen to make the initial adjustments. Afterward, less specular objects may be examined. Follow these steps to arrange the microscope for Köhler illumination, and refer to Figure 10-3.

Initial Adjustment

1. While looking at the specimen from about stage level, bring the specimen and objective close to one another, using the coarse focus adjustment knob.
2. While looking in the microscope eyepiece, carefully focus up with the coarse adjustment until the specimen is in focus. It is easier to spot a moving specimen, so if you have a rotating stage it is a good idea to rotate it back and forth in a short arc. Watch for small dust particles or surface irregularities to come into focus first. Touch up the focus by adjusting the fine focus.
3. With the specimen in focus and illumina-

tion adequate, center objectives that are in centerable nosepiece mounts. This step is for microscopes with rotating stages; microscopes with fixed stages do not require this operation.

4. If the illumination is adequate, adjust the binocular tubes when a binocular or trinocular head is being used.

5. With the specimen focused, the objective centered, and the binocular head properly adjusted for both eyes, you are ready to start the most critical steps in achieving Köhler illumination. Close the field diaphragm. This will be the closer of the two levers or knurled collars to the microscope objective. The lens between the field diaphragm and the body tube, together with the objective itself, acts as a condenser to focus a sharp image of the field diaphragm in the plane of the specimen. Both the specimen and the leaves of the diaphragm will be seen in sharp focus. The exact degree of sharpness of the field diaphragm image will depend to some extent on the degree of correction of the objective. The field diaphragm and the lens following it are usually fixed at the factory and are, therefore, nonfocusable (unlike transmitted-light microscopes in which the substage condenser is adjusted to bring the image of the field diaphragm in focus in the plane of the specimen).

If the lens between the field diaphragm and the body tube is adjustable, or if the distance between the field diaphragm and the lens can be varied, make these adjustments so that a sharp, centered image of the field diaphragm lies superimposed on the sharply focused specimen. There may be some way of centering the image of the field diaphragm in the field of view; if so, make that adjustment. The centering of the field diaphragm may be easier to accomplish if it is opened until it is near the edge of the field of view. When centering and focusing are completed, open the field diaphragm until it is just outside the field of view. This is the normal position of the field diaphragm for both visual use and photomicrography. With some specimens, contrast is increased through flare reduction by closing down the field diaphragm. Remember to open the field diaphragm before making a photomicrograph—at least to just beyond the format indicating lines. When the specimen and field diaphragm are both in focus, their image will also be in focus at the intermediate image plane for subsequent secondary magnification by the eyepiece, whose diaphragm also lies in the intermediate image plane.

Adjust the Lamp

6. In the next step you must focus a real image of the lamp filament in the plane of the aperture diaphragm, using the lamp condenser (collector). To do this, adjust the separation between the filament and the lamp condenser. The lamp condenser may be adjustable, but it is far more usual to find the lamp condenser fixed and the lamp adjustable, usually by sliding the lamp holder back and forth in its cylindrical housing. In either case, adjust the distance between the filament and the aperture diaphragm until the image of the filament is projected in sharp focus on the (temporarily closed) aperture diaphragm. If this adjustment is made, an image of the filament and the aperture diaphragm will be in focus at the objective back focal plane.

So far, we have been describing the setup for Köhler illumination when using the plane-glass reflector in the vertical illuminator, but what if you decide to use the prism reflector? In this case, arrange the lamp focus and alignment so that the entire filament image lies on the face of the prism that is perpendicular to the light path from the lamp. If you leave the lamp adjusted for the plane-glass reflector and then introduce the prism, you lose about half of the incident illumination.

Adjust the Aperture Diaphragm

7. After the objective back focal plane is fully illuminated with a centered, focused filament, close down the aperture diaphragm (the lever or knurled collar farther away from the body tube) while viewing the objective back focal plane until the diaphragm just comes into the field of view superimposed on the focused filament. If this adjustment is correctly made, the images of the filament and aperture diaphragm will also appear at the eyepiece exit pupil where the lens of the eye or camera is placed.

8. Replace the eyepiece (or take out the Bertand lens or phase telescope) and observe the field of view. The image should now be brightly and evenly illuminated with good depth of field and resolving power. The metallographic microscope is now ready to make high quality photomicrographs.

It is exceedingly important to note, however, that as objectives are changed, the adjustment of both the aperture and field diaphragms must be altered.

Once the specimen is correctly focused and illuminated, photography can proceed as with transmitted light.

Photography

The interference colors of some metallographic specimens will photograph disappointingly on some color films. If you encounter this situation, try a film of different speed, alternate color balance (with correction filter), or different dye system.

The distortion (usually pincushion) of many objectives is not noticed until rectilinear objects, such as microcircuits, are photographed. This distortion is inherent in the lenses so that no optical or photographic adjustment can cure the fault.

Chapter Eleven

RECORDING MOTION THROUGH THE MICROSCOPE

At the outset, every photomicrographer is encouraged to try *cinemicrography*,* or the making of motion pictures through the microscope. A quick survey of microscopists shows that cinemicrography is regarded as forbidding because it is too difficult and complex (it is neither) and because it is too expensive (it need not be). Once still photomicrography is mastered so that good, well-illuminated visual images can be consistently produced and recorded with still cameras, the transition to cinemicrography is very simple. Reduced to basic terms, the differences between still photomicrography and cinemicrography are framing rate and lighting. Instead of one frame being exposed at a time in relatively very slow sequence, the motion-picture camera allows a relatively rapid sequence of frames to be exposed in succession.

The mechanics may seem imposing at first, but all devices we are not familiar with seem that way. The modern motion-picture camera is hardly more complex in operation than a still camera. Many 8 mm and 16 mm motion-picture cameras—especially those designed for amateur photography—are simpler to use than many 35 mm still cameras that professionals use.

A good way to get started is to borrow a friend's super 8 motion-picture camera. Biomedical and industrial photomicrographers may wish to rent professional equipment for weekend use, just to have a go at it. Lighting may become tricky if a variable-speed motor is available to change the framing rate, but this is not a problem when it is understood. Motion-picture film and processing are surprisingly inexpensive. The results will ultimately always come down to technique, skill, and interpretation, rather than film cost and equipment complexity.

Cinemicrography

Cinemicrography involves the adaptation of a motion-picture camera to a compound microscope to record images of moving specimens. Three basic techniques are included: (1) recording extremely slow motion by

*Also cine*photo*micrography.

making a long series of separate exposures at predetermined intervals (time lapse) and projecting the result at normal speed, thereby greatly increasing the apparent rate of motion; (2) photographing normal motion at the speed of occurrence and projecting the film at the same or similar speed; and (3) photographing motion at a high rate of speed and projecting the film at a normal speed to produce a slow-motion effect.

Both camera speed and projection speed are expressed as the number of film frames per second (fps). For motion pictures with sound, normal speed is 24 fps. Normal speed for silent motion pictures is often 18 fps although 24 fps may be used also. Ultimate projection speed should always be considered in planning cinemicrography.

Microscope

The general arrangement of equipment for making motion pictures through a microscope is similar to that for making a still photomicrograph. The camera is placed so that the image ordinarily projected to the eye is projected to the film.

In cinemicrography, the convenience of being able to see the field of action (either during or immediately preceding the making of the motion picture) is so important that it is almost a necessity. This is usually accomplished by a beam splitter that divides the beam of light forming the image so that a small portion of it is carried to the observer's eye and the remainder to the film in the camera.

As with visual observation, one needs a suitable light source. In this case, it must be of sufficient intensity to produce satisfactory exposure on the motion-picture film. For still photomicrography, intensity of light is not as important as it is for motion pictures. If the light is dim, time exposures can be made for satisfactory still pictures of immobile subjects. In making motion pictures at high camera speeds (increased frame rates), short exposure times are common. Hence, the intensity of the illumination must be raised over that usually employed for visual observation or still photomicrography.

A microscope with built-in illumination may not be suitable for all types of cine-

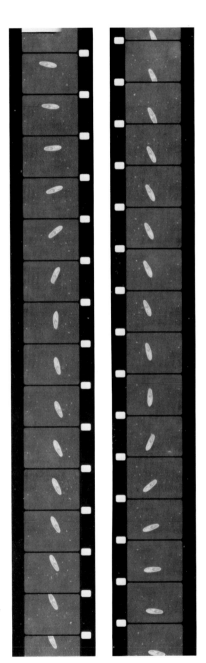

Fig. 11–1

MOVEMENT—*With a 16-mm motion-picture camera, movement of a living specimen can be recorded.*

micrography because the light source may be of insufficient intensity. Very often, a microscope equipped with a substage mirror is best since this feature allows a variety of light sources to be used in detached illuminators.

The effective use of any microscope requires that the illumination and the microscope optics be correctly aligned and adjusted. Köhler illumination (see page   ), when followed faithfully, will produce the best image quality and the brightest illumination of the specimen to be photographed.

Camera

While most motion-picture cameras can be adapted for cinemicrography, superior results will be obtained if one of the better cameras is chosen. Professional-type motion-picture cameras are usually best since they are often equipped to make single-frame exposures for use in time-lapse photography. Ordinarily they can also be operated at various camera speeds (fps) for both normal and slow-motion results. For less critical work, *any* motion-picture camera can be used.

A suitable camera is designed to accommodate either a detachable film magazine (to facilitate reloading) or a long length of film (usually 30 metres or 100 feet). The camera should be driven either by an electric motor or by a spring of sufficient capacity to allow exposure of a useful length of film at one winding. A variable shutter is an advantage in controlling exposure time.

Cameras with fixed lenses can be used over a microscope with reasonably good results, but they entail certain precautions. The best image quality is obtained when the camera is placed so that the eyepoint (Ramsden disk) of the eyepiece occurs at or very near the front surface of the camera lens. This is not always possible since some camera lenses are recessed in the mount so that the lens surface is too far from the eyepoint. The result is a vignetting of the microscope image in the film plane; the frame is not covered completely, and only a very small circle appears in the center of the film frame. If the front of the camera lens mount can be removed, this condition is often alleviated, since the correct position can be achieved.

If a lens of normal focal length (25 mm for 16 mm cameras) is used and correctly placed as indicated, the film plane is so near the eyepoint that the size of the microscope image may be approaching, or even less than, the frame size. Then all, or almost all, of the microscope field is recorded instead of just the central area, which is the area of best image definition. Lenses of longer focal length (such as 50 mm and 63 mm) are better

in this respect because of their narrower angle of view.

For ordinary purposes, magnification in the microscope can be found by multiplying the objective magnifying power by the ocular magnifying power. Thus, a 10X objective and 10X ocular will furnish 100X. When the image is projected at 250 mm (10 inches) above the ocular, this magnification is reproduced. However, when a camera with lens attached is placed correctly over the microscope, the distance from eyepoint to film is approximately the focal distance of the lens, and magnification is reduced proportionately. A 50 mm (2-inch) lens will result in an image only 1/5 (50/250) the size of the image seen in the microscope. Thus, an indicated magnification of 100X is reduced to only 20X as recorded at the film plane. Lenses of other focal lengths change magnification proportionately. These factors are not necessarily disadvantageous if the entire film frame is covered, but they should be considered when correct recording of magnification is important.

If a microscope is focused visually and a camera (with attached lens) is placed in the correct position, the distance setting on the camera lens should be placed at infinity. The microscope image will be in focus in the camera. Those who cannot focus their eyes at infinity should make test runs at various distance settings of the camera lens in order to find the virtual distance at which they do focus. The lens diaphragm should be set at or near its wide-open position. A small lens opening may vignette the image. The diaphragm of the camera lens *does not* control exposure.

Beam Splitter

A beam splitter with observation eyepiece is essential between the camera and the microscope, not only for viewing and focusing the image during photography but also for the critical centering and illumination adjustments necessary for high-quality work.* In order to photograph rapidly moving organisms, it is necessary to know that they are really in the field area. Photographing very slow motion is less exacting since the subject will not leave the area during photography. However, you cannot be sure that the image remains in correct focus. A beam-splitting observation eyepiece gives both correct focus and recording of pertinent detail. Some beam splitters require that a lens be in place on the camera, whereas others attach directly to the camera with the lens removed. Beam splitters

*Through-the-lens viewing available with professional cine cameras can provide a means to follow action, but critical focusing may be difficult.

are manufactured and sold by several firms for both still and motion-picture photography with the microscope.

Techniques

Time Lapse

The equipment for time-lapse studies through a microscope consists of a basic cinemicrographic setup plus a suitable controller for making single-frame exposures at any desired predetermined interval. Several commercial controllers are available. Basically, these consist of an electric timer, its associated power-control components, and a solenoid or other device for automatically actuating the exposure release of the motion-picture camera. For time intervals of more than a few seconds between exposures, provision is often made within the control unit for turning on the illumination just before the exposure and for extinguishing it after exposure. Thus, because the illuminator is not required to operate for long periods of time, heating and burnout problems are reduced. Once adjusted and operating properly, the apparatus can be left unattended; it will proceed automatically to record the action overnight, or even for weeks if needed. Of course, it is necessary to wind the spring motor of the camera occasionally (if an electric-drive motor is not used), to check the bulb in the illuminator periodically, and also to check focus and composition.

Time-lapse techniques are particularly useful in cinemicrography. They permit the relatively slow growth phenomena of microscopic plants and animals to be recorded completely over a period of hours, days, or even weeks. When the film is projected, the whole cycle can be viewed in only a few minutes. Changes that are so slow and subtle as to be difficult to discern visually become quite evident at an accelerated rate. Chemical and physical occurences, such as crystal growth, can also be projected with an alteration of the time base for convenient observation.

A widespread use of time-lapse photomicrography is for recording the growth and change in cells in tissue culture using a phase-contrast or interference-contrast microscope.

One of the principal problems in time-lapse photography is determining an appropriate time interval between exposures. An accepted technique is to make a visual study of the complete action to be recorded, noting the time involved from beginning to end. For example, suppose that the complete action took place in 1 hour. This action is to be recorded in a 16 mm camera on a 100-foot roll of film. There are 40 frames in 1 foot of 16 mm film, or 4000 frames in a 100-foot roll. One hour's time equals 3600 seconds; so at 1 frame per second, the complete action could be recorded on the available roll of film. If the film is projected at sound speed (24 frames per second), the action will be speeded up 24 times. Total time for projection of the 1-hour action is only 150 seconds, or 2.5 minutes. If somewhat slower projection is desired, the projector can be run at silent speed (18 frames per second). The projection time would be 3.75 minutes.

With 72 frames per foot, a super 8 film has 3600 frames. At a 1 frame per second exposure rate, the 50-foot film supply will last 1 hour. Total projection time at 18 frames per second will be 3.33 minutes.

Intervals in Time-Lapse Cinemicrography

Action Time	Interval*
15 min	4 frames per sec
30 min	2 frames per sec
60 min (1 hr)	1 frame per sec
2 hrs	1 frame each 2 sec
4 hrs	1 frame each 4 sec
8 hrs	1 frame each 8 sec
24 hrs	1 frame each 24 sec

*Based on 3600 frames—all of a 50-foot super 8 film or most of a 100-foot 16 mm film.

Time intervals of longer than 1 second between frames are indicated, of course, if the total action time is in excess of 1 hour. Conversely, if the action takes only a few minutes, the time interval will be much shorter.

Normal Speed

Normal-speed cinemicrography is analogous to conventional motion-picture photography. If the camera operates at a given rate of speed and the film is projected at that same rate, the subject motion will appear the same as that seen in the microscope. The camera can be run at various speeds, depending on the actual speed of motion of the specimen. The film is then projected at a speed that will duplicate or approximate the motion of the specimen. Normal camera speeds can be 8, 18, 24, or 32 frames per second.

Slow Motion

Any organisms that move very fast may require slow-motion cinemicrography in order to facilitate a visual analysis of the motion itself or to allow an examination of subject details.

Exposure Times (sec)* for Variable Shutters

Camera Speed (frames per sec)	Shutter Opening							
	200°	180°	160°	100°	90°	80°	50°	45°
8	1/14	1/16	1/18	1/28	1/32	1/36	1/56	1/64
18	1/28	1/32	1/36	1/56	1/64	1/72	1/112	1/128
24	1/43	1/48	1/54	1/86	1/96	1/108	1/172	1/192
32	1/56	1/64	1/72	1/112	1/128	1/144	1/224	1/256
48	1/86	1/96	1/108	1/172	1/192	1/216	1/354	1/384
64	1/112	1/128	1/144	1/224	1/256	1/288	1/488	1/512
128	1/224	1/256	1/288	1/448	1/512	1/576	1/896	1/1024

*Exposure time for shutter openings not shown can be computed with the following formula:

$$\text{Exposure time} = \frac{\text{Shutter opening (in degrees)}}{\text{fps} \times 360°}$$

Whenever the camera's taking speed exceeds the projection speed, the recorded action appears slower than that seen in the microscope. Camera speed may be 32, 48, 64, 125, or even up to 200 frames per second with conventional types of motion-picture cameras. At normal projection rate (18 or 24 fps) of films made at any of these camera speeds, the image moves more slowly than the original action. This permits detailed study of subject motion when the original motion is too rapid for direct visual analysis. Making motion pictures at high speeds requires the use of a high-intensity light source or a high-speed film or both.

Illumination

In order to use any microscope and light source efficiently, it is necessary to follow a definite technique of illumination. Köhler illumination has been widely accepted as the most useful means of adjusting a microscope and obtaining efficient illumination of the specimen. The details of Köhler illumination can be found on page 28.

A separate illuminator (not built in) allows the use of a variety of light sources, with highly efficient illumination. Many modern microscopes contain built-in illumination, often utilizing a modified Köhler system. However, such self-contained illumination may be of relatively low intensity and, therefore, not suitable for most motion-picture work.

Light Sources

The *tungsten lamps* and, especially, the *tungsten-halogen* lamps mentioned earlier (page 24) can all be used in cinemicrography systems such as time lapse, in which the light may be turned on and off at spaced intervals. More intense light sources such as xenon-arc lamps, carbon arcs, or mercury-vapor lamps may be needed for other techniques.

An *electronic flash* lamp is particularly useful in time-lapse cinemicrography such as in tissue-culture work. The extremely short (1/1000 second or less) burst of high-intensity illumination is of great advantage in stopping motion and in preventing heat damage to the specimen. An auxiliary steady light source of low intensity is necessary when you adjust the microscope, locate the field, or focus the image between exposures.

Since capacitors in a power source have to recharge between exposures, the recharging time must be less than the time-lapse interval. If the recharging time is greater than the time-lapse interval, the lamp will not flash with full brightness. Repetitive-flash units—those that are specially designed to be flashed regularly at full power—are indispensable for many cine applications.

Exposure Control

Time

Exposure time can be varied in motion-picture work by using a variable-opening shutter, by changing camera speed (frames per second), or by changing the time during which the shutter remains open.

Some professional cameras contain rotating disk shutters in front of the film plane. The open segment of this disk is variable from fully open (170 degrees, usually) to one-quarter open (often with intermediate settings). At 8 frames per second and with the shutter fully open, the exposure time is about 1/30 second. At one-half open, the exposure time will be 1/60 second; at one-quarter open, it will be 1/120 second. When other camera speeds are used, the partial openings will give proportionately less exposure times. This variable shutter thus allows some control of exposure by changing exposure time.

Exposure time can also be varied by changing camera speed. Remember that this causes an apparent change in rate of subject motion when the film is projected at a normal speed.

In time-lapse cinemicrography, it is often desirable to vary exposure time and to allow image brightness to remain constant. Some time-lapse controllers actuate the camera shutter and can also hold it open for any length of time. It is also possible in time-lapse photography to utilize a separate variable electric shutter in the light beam. In this case, the control unit automatically opens the camera shutter and actuates the electric shutter for a preset time.

Light Intensity

The intensity of light reaching the film is the other factor that controls exposure. It is fairly easy to decrease the intensity of light by means of neutral densities as previously stated. To increase the intensity of light, you must change electrical conditions by raising voltage or amperage. This can be done, within limits, for black-and-white films but it should not be used with color films since illumination color quality (color temperature) will be altered. Generally, when image brightness in the observation eyepiece is too low, it is necessary to resort to a light source of higher intensity to obtain a bright image on the film.

Glossary of Terms Used in Photomicrography

Aberration—Any inherent deficiency of a lens or optical system which is responsible for imperfections in the shape or sharpness of the image as a result of failure to image a point or a straight line as such or an angle as an equal angle. The various forms of aberration can be reduced but not completely eliminated by assembling a lens from a number of different elements of compensating characteristics.

Achromat—A lens which brings light from two parts of the spectrum (strictly speaking, two specified wavelengths) to the same focus. Modern achromatic lenses bring the blue and red regions of the spectrum to the same focus, thus reducing chromatic aberration.

Anisotropic—Exhibiting different properties when measured along different axes. A transparent, anisotropic material such as a crystal of calcite, possesses different refractive indices in different directions through its mass (birefringence) and polarizes transmitted light.

Aperture—The lens opening through which light enters an optical instrument. The area of this opening is sometimes adjustable by an iris diaphragm or stop.

Apochromat—Lens which brings three chosen colors to the same focus. Apochromats used for microscope objectives are corrected for chromatic aberration for red, green, and blue wavelengths.

Betrand Lens—Removable positive lens which can be fitted above the objective of the microscope to form an image of the objective's rear focal plane in the front focal plane of the eyepiece. It enables the *exit pupil* to be observed without removing the eyepiece when setting up Köhler illumination.

Birefringence—The refraction of light in two slightly different directions to form two rays. Polarizing elements that contain crystals oriented in one direction are birefringent.

Chromatic Aberration—Faults in the performance of lenses due to light of different colors coming to different planes of focus, the blue image being nearest the lens, or yielding images of different magnification.

Compensating Eyepiece—An eyepiece corrected primarily for use with apochromatic objectives, eliminating the color fringes found when ordinary eyepieces are used with such objectives.

Condenser—An optical assembly which concentrates light from a source. It provides an evenly illuminated image of the source used in photomicrography, projection of slides, and printing negatives.

Contrast—The ratio of the amount of light transmitted or reflected by the most transparent and most opaque areas of an image.

Curvature of Field—A lens aberration in which images can be sharply focused only on a curved surface.

Depth of Field—The region in front of and behind the focused distance within the subject in which object points still produce an image of acceptable sharpness.

Depth of Focus—Tolerance in the positioning of the image plane of a lens within which the lens forms an acceptably sharp image of an object at a given distance. In focusing, it represents, in effect, a focusing latitude.

Distortion—Aberration of a lens which causes the image to appear misshapen and deformed due to a gradual increase or decrease in magnification from the center to the edge of an image.

Empty Magnification—Magnification achieved by increasing the size of the image but not detail. This is caused by limited resolving power of the optical system.

Interference Microscopy—Technique by which the illumination is divided into two beams, one of which passes through the transparent subject matter. The second beam is passed around, rather than through, the specimen. Recombination of the two beams results in interference patterns corresponding to the local thickness variations in the specimen.

Köhler Illumination—A method of brightfield illumination used in photomicrography and cinemicrography. A lamp collector lens focuses an image of the light source on the aperture diaphragm; and the microscope collector focuses the field diaphragm in the plane of the specimen. This technique provides a uniformly illuminated field from nonuniform light sources such as coiled filament lamps.

Nanometre (nm)—SI unit of measure for electromagnetic radiation. Equals 10^{-9} metre. Visible light wavelengths range from 400 to 700 nanometres.

Numerical Aperture (NA)—A method of specifying the relative aperture of an objective lens and its resolving power. The numerical aperture value refers to the angle of the cone of light emitted by the condenser and accepted by the objective of the microscope. The formula is NA = d times the sine of U, where D is the refractive index of the space between the objective and the specimen, and U is one-half the angular cone of illumination required to fill the aperture of the front lens of the objective. About 1,000 times the numerical aperture indicates the approximate limit of useful magnification of an objective.

Parfocal—The property of several lenses of different focal lengths being in focus at the same position of the focusing knob.

Phase Contrast Microscopy—A technique for revealing the structural features of microscopic transparent objects whose varying but invisible differences in thickness result in varying differences in the phase of transmitted light. These phase differences are converted to visible intensity differences when part of the transmitted light has its optical path changed by about 1/3 wavelength.

Polarizer—A transparent material which absorbs from light passing through it all vibrations except those in a single plane.

Polarizing Element—A material which transmits light polarized in one particular plane. Two such elements can be used to distinguish crystals of different types in photomicrography. If visually identical crystals differ in their birefringence, they will show differences in color and tone patterns if one polarizing filter is below and the other (which is capable of being rotated) is above the specimen.

Refraction—Change in direction of a ray of light passing from one transparent medium into another of a different optical density, e.g., from air into glass.

Refractive Index—The ratio of the speed of light in a vacuum to its speed in some other medium. This ratio determines how much light rays are bent.

Resolving Power—The ability of a lens to distinguish fine detail in the structure of a specimen. This ability is assessed by counting the number of closely spaced objects or line pairs that can be recognized as separate in the final image. One formula is resolving power equals wavelength of light divided by two times the numerical aperture.

Spherical Aberration—A lens defect in which light rays passing through the outer regions of a lens converge and cross the lens axis nearer to the lens than rays passing through the central part of the lens. An object point at the lens axis is recorded as a disc of light.

Selected References

General Microscopy

Bennett, A. H., H. Jupnik, H. Osterberg, and O. W. Richards (1951). *Phase Microscopy.* John Wiley, New York.

Burrells, W. (1964). *Industrial Microscopy in Practice.* Fountain Press, London.

Chamot, E. M., and C. W. Mason. *Handbook of Chemical Microscopy.* Vol. 1, 3rd Ed., 1958; Vol. 2, 2nd Ed., 1940. John Wiley, New York.

Gude, W. D. (1968). *Autoradigraphic Techniques.* Prentice-Hall, Englewood Cliffs, New Jersey.

Hallimond, A. F. (1970). *The Polarizing Microscope,* 3rd Ed. Vickers, Ltd., London.

Martin, L. C. (1966). *The Theory of The Microscope.* American Elsevier, New York.

Photomicrography

Bergner, J., E. Gelbke, and W. Mehliss (1966). *Practical Photomicrography.* Focal Press, London and New York (Amphoto).

Gander, R. (1969). *Photomicrographic Technique for Medical and Biological Scientists.* Hafner, New York.

Hassett, G. F., and T. P. Hurtgen. (1972). "Photomicrography in Dentistry," *Dental Radiography and Photography,* 45, 51-53, 59.

Kloserych, S. (1964). "Photomicrography—Exposure Determination," *J. Biol. Phot. Assoc., 32* 133-147.

Lawson, D. F. (1972). *Photomicrography.* Academic Press, New York.

Loveland, R. P. (1970). *Photomicrography; A Comprehensive Treatire,* 2 vols. John Wiley, New York.

Loveland, R. P. (1971). "Characteristics and Choice of Photographic Materials for Photomicrography," *The Microscope 19,* 177-203.

Michel, K. (1962). Die wissenschaffliche und angewandte Photographie, 2nd Ed., Springer, Vienna.

Mollring, F. K. (1966). "Methods of Optical Adaptation of the Movie Camera to The Microscope." *Microscopy 30,* 181-191.

Rose, G. G. (1963). *Cinemicrography In Cell Biology.* Academic Press, New York.

Shillaber, C. P. (1944). *Photomicrography In Theory and Practice.* John Wiley, New York.

Spinell, B. M. (1961). "Simplified 365mm Photomicrography With Improved Results," *J. Biol. Phot. Assoc. 29,* 145-152.

Szabo, D. (1967). *Medical Colour Photomicrography.* Akademiai Kiado, Budapest.

Traber, H. A. (1971). *The Microscope As a Camera.* Focal Press, London and New York (Amphoto).

Walker, M. I. (1971). *Amateur Photomicrography.* Focal Press, London and New York (Amphoto).

Metallography

Chalmers, B., and A. S. Quarrel. (1960). *Physical Examination of Metals: Vol. 1 Optical Methods.* Arnold, London.

Gifkins, R. C. (1970). *Optical Microscopy of Metals.* American Elsenier, New York.

Greaves, R. H., and H. Wrighton (1956). *Practical Microscopical Metallography,* 4th Ed. Chapman and Hall, London.

Kehl, George L. (1949). *The Principles of Metallographic Laboratory Practice,* 3rd Ed., McGraw-Hill, New York.

Kehl, George L. (1964). *Metallography.* American Society for Metals, Metals Park, Ohio.

Modin, H., and S. Modin (1973). *Metallurgical Microscopy.* John Wiley, New York.

Nutt, M. C. (1968). *Principles of Modern Metallurgy.* Charles E. Merrill, Columbus, Ohio.

Samuels, L. E. (1967). *Metallographic Polishing By Mechanical Methods.* Pitman, London.

Smithell, C. J. (1967). *Metals Reference Book,* Vol. 1.

Photographs, photomicrographs in this book
As acknowledged with individual photograph. The following are not otherwise acknowledged:
John Gustav Delly—inside covers, 26, 30, 32, 33, 34, 35, 36, 37, 39.
Bruce W. Grant—57 (Fig. 5-6)
M. L. Scott—56 (Fig. 5-3, 5-4), 57 (Fig. 5-5), 63 (Fig. 7-1)

Appendix I
SOME SOURCES OF PREPARED MICROSLIDES

Carolina Biological Supply Co.
York Road
Burlington, NC 27215

Ripon Microslides, Inc.
P.O. Box 262
Ripon, WI 54971

Triarch, Inc.
P.O. Box 98
Ripon, WI 54971

Turtox/Cambosco
Macmillan Science Co., Inc.
8200 S. Hoyne Ave.
Chicago, IL 60620

Wards Natural Science Establishment, Inc.
P.O. Box 1712
Rochester, NY 14603
or
P.O. Box 1749
Monterey, CA 93940

Appendix II
FILM CHOICE FOR PHOTOMICROGRAPHY

Use these lists with the decision chart on page 92.

List A
Black-and-White Films

KODAK Film	Sheet Film or Roll Sizes	Description	Exposure Index	Exposure Compensation (Time) and Development Adjustment* for Reciprocity Characteristics					
				1/1000	1/100	1/10	1	10	100
PANATOMIC-X	135	Slow-speed panchromatic film	32	No exposure change +30% development	No exposure change +20% development	No exposure change +20% development	Increase exposure time 2X +10% development	Increase exposure time 5X Normal development	Increase exposure time 12X −10% development
PLUS-X Pan	135	Medium-speed panchromatic film	125						
PLUS-X Pan Professional	4147 (ESTAR Thick Base)	Medium-speed panchromatic film	125						
TRI-X Pan	135	High-speed panchromatic film	400						
TRI-X Pan Professional	4164 (ESTAR Thick Base)	High-speed panchromatic film	320						

*Development adjustments *include* a 20% development increase recommended for black-and-white films exposed through the microscope.

List B
KODAK
EKTACHROME Films
Process E-6

KODAK Film	Sheet Film or Roll Sizes	Description	Film Balance	Daylight Exposure Index, Filter	3200 K Tungsten Exposure Index, Filter	Exposure Compensation (Stops) and Filter for Reciprocity Characteristics When Used With Type of Light for Which Film is Balanced					
						Exposure Time in Seconds					
						1/1000	1/100	1/10	1	10	100
EKTACHROME 64 Professional (Daylight)	6117 135, 120	Professional daylight-balanced color transparency film	Daylight	64 NF	16 80A	None NF	None NF	None NF	+1 CC15B	+1½ CC20B	NR
EKTACHROME 200 Professional (Daylight)	135, 120	Fast professional daylight-balanced color transparency film	Daylight	200 NF	50 80A	None NF	None NF	None NF	+½ CC10R	NR	NR
EKTACHROME 50 Professional (Tungsten)	6118 135, 120	Professional tungsten-balanced color transparency film	Tungsten (3200 K)	32 85B (1/60 sec)	50 NF (1/10 sec)	NR	None NF	None NF	None NF	+1 CC10B	NR
EKTACHROME 160 Professional (Tungsten)	135, 120	Fast professional tungsten-balanced color transparency film	Tungsten (3200 K)	100 85B (1/60 sec)	160 NF (1/30 sec)	None NF	None NF	None NF	+½ CC10R	+1 CC15R	NR
EKTACHROME 64 (Daylight)	110, 135, 126, 127	General-purpose medium-speed daylight-balanced color transparency film	Daylight	64 NF	16 80A	None NF	None NF	None NF	+1 CC15B	+1½ CC20B	NR
EKTACHROME 200 (Daylight)	126, 135	Fast daylight-balanced color transparency film for general use	Daylight	200 NF	50 80A	None NF	None NF	None NF	+½ CC10R	NR	NR
EKTACHROME 160 (Tungsten)	135	Fast tungsten-balanced color transparency film for general use	Tungsten (3200 K)	100 85B	160 NF	None NF	None NF	None NF	+½ CC10R	+1 CC15R	NR
EKTACHROME 400 (Daylight)	135, 120	Highest speed color transparency film made by Kodak	Daylight	400 NF	100 80A	None NF	None NF	None NF	+½ NF	+1½ CC10C	+2½ CC10C

NF: No Filter   NR: Not Recommended

List C
KODACHROME Films
Process K-14

KODAK Film	Sheet Film or Roll Sizes	Description	Film Balance	Daylight Exposure Index, Filter	3200 K Tungsten Exposure Index, Filter	Exposure Compensation (Stops) and Filter for Reciprocity Characteristics When Used With Type of Light for Which Film is Balanced					
						Exposure Time in Seconds					
						1/1000	1/100	1/10	1	10	100
KODACHROME 25 (Daylight)	135, 828	General-purpose moderate-speed daylight-balanced color transparency film	Daylight	25 NF	6 80A	None NF	None NF	None NF	+1 CC10M	+1½ CC10M	+2½ CC10M
KODACHROME 64 (Daylight)	126, 110, 135	General-purpose medium-speed daylight-balanced color transparency film	Daylight	64 NF	16 80A	None NF	None NF	None NF	+1 CC10R	NR	NR
KODACHROME 40 (Type A)	135	Moderate-speed photolamp-balanced color transparency film	Photolamp (3400 K)	25 85	32 82A	None NF	None NF	None NF	+½ NF	NR	NR

NF: No Filter   NR: Not Recommended

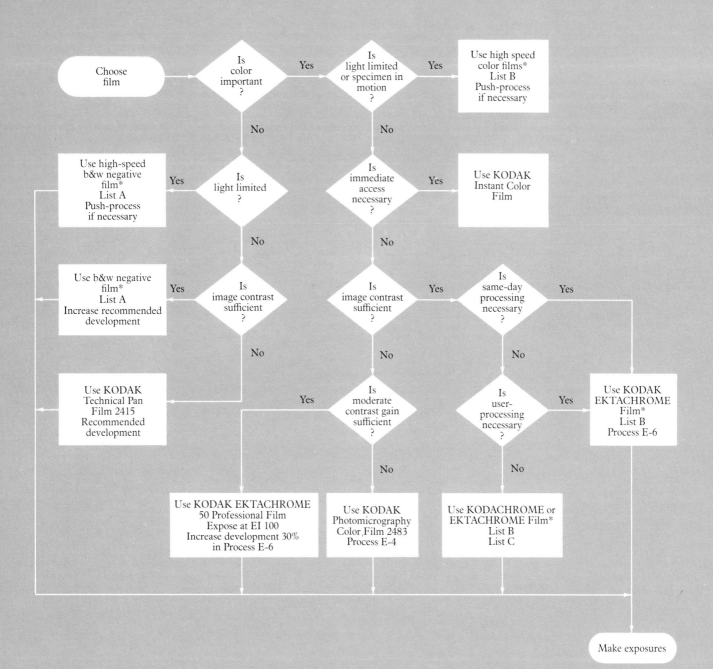

Choose film

Is color important?
— Yes → Is light limited or specimen in motion? — Yes → Use high speed color films* List B Push-process if necessary
— No ↓

Is light limited?
— Yes → Use high-speed b&w negative film* List A Push-process if necessary
— No ↓

Is image contrast sufficient?
— Yes → Use b&w negative film* List A Increase recommended development
— No ↓ → Use KODAK Technical Pan Film 2415 Recommended development

Is light limited or specimen in motion? — No ↓
Is immediate access necessary?
— Yes → Use KODAK Instant Color Film
— No ↓

Is image contrast sufficient?
— Yes → Is same-day processing necessary?
— No ↓

Is moderate contrast gain sufficient?
— Yes → Use KODAK EKTACHROME 50 Professional Film Expose at EI 100 Increase development 30% in Process E-6
— No → Use KODAK Photomicrography Color Film 2483 Process E-4

Is same-day processing necessary?
— Yes → Use KODAK EKTACHROME Film* List B Process E-6
— No ↓

Is user-processing necessary?
— Yes → Use KODAK EKTACHROME Film* List B Process E-6
— No → Use KODACHROME or EKTACHROME Film* List B List C

Make exposures

*For finer grain and higher resolution, use slowest film that level of illumination permits.

Appendix III
REVERSAL COLOR FILMS FOR PHOTOMICROGRAPHY

KODAK EKTACHROME Films (Process E-6)

KODAK EKTACHROME 50 Professional Film (Tungsten)—For general-purpose photomicrography on microscopes with tungsten-halogen or ordinary tungsten lamps. Has favorable reciprocity characteristics for longer exposures, when required. Can be exposed and processed for an effective exposure increase of one stop (ASA 100) to gain more contrast. Responds well to the use of a didymium filter for improved rendition of H&E stain and other stains containing eosin or fuchsin.

KODAK EKTACHROME 64 Professional Film (Daylight) or KODAK EKTACHROME 64 Film (Daylight)—Balanced for 5500 K, making it usable with microscopes having xenon lamps or electronic flash.

KODAK EKTACHROME 160 Professional Film (Tungsten)—Well suited to the photography of specimens (particularly three-dimensional ones) by reflected light in the stereo microscope. Can be processed for an effective exposure index of 320 to allow shorter exposures.

KODAK EKTACHROME 200 Professional Film (Daylight) or KODAK EKTACHROME 200 Film (Daylight) and KODAK EKTACHROME 400 Film (Daylight)—Excellent choice for fluorescence micrography, especially with weak-emitting specimens, or with fluorochromes which fade (quench) rapidly. When processed for normal film speeds or for higher effective speeds, each of the films fills the need for detection of weak or fleeting emission.

The EKTACHROME Films for Process E-6 display better response when used with a didymium filter to improve histological stain rendition than did the older EKTACHROME Films.

KODAK EKTACHROME Professional Films (Process E-6) are released for sale "on-aim" for color balance and are packaged with specific exposure information that is determined for each individual emulsion batch. The films are stored and distributed by Kodak under refrigerated conditions and then similarly cared for by the professional stockhouse dealers who stock these films. Thus, to maintain the speed and color balance characteristics that are incorporated in manufacture, the user also should plan to provide refrigerated storage for the film. KODAK EKTACHROME Professional Films are designed for prompt processing (within a week) so that the optimum color balance does not shift significantly after exposures are made.

EKTACHROME Films without the professional designation are intended for general picture-taking. These films are specifically designed and release-tested for room-temperature storage and for casual photography where there may be a considerable time between exposures or between exposure and processing of the film. The aging effects which are inevitable with such use are taken into account in manufacture. For example, if a given film is known to shift toward yellow-green as it ages, it will be manufactured with its color balance shifted in the opposite direction (toward blue-magenta) to compensate. Such a film will lose its excess blue-magenta balance during the time it is being shipped and in stock on the dealer's shelf. From then on, there will be a considerable time during which the color balance of the film is near its optimum point. This is the time during which the film is most likely to be used and processed.

Processing for Increased Film Speed and Contrast in Process E-6

For most tasks, EKTACHROME Films (Process E-6) will have adequate speed and contrast when exposed and processed as recommended. Occasionally, however, the nature of the specimen, the optical techniques used, or photomicrographer's preference may make it desirable to increase the effective speed or contrast of the film. Do this by first exposing the film using a higher than normal exposure index, e.g., EI 100 instead of EI 50 and then increasing the duration of the first development step. The user can make the appropriate processing modification by following instructions packaged with Process E-6 chemicals or by instructing a processing laboratory to do so. Kodak laboratories will provide processing for a 2X increase in the nominal film speed when the film is mailed in a KODAK Special Processing Envelope, ESP-1. Users and independent processors can achieve 2X and, in some cases, even higher effective speeds.

In general, the two lower speed films (ASA 50 and 64) show greater gains in contrast when processed to higher speeds than do those with inherently higher speed. All of these films lose very slightly in maximum density when processed for a one-stop speed increase. This is immaterial in brightfield micrography, since little or none of the specimen is reproduced at that level of blackness. Even in darkfield or fluorescence work, slight loss in D-max is an acceptable trade-off for the gain achieved in speed. When the film is processed to gain two stops in speed, the loss of maximum density is more noticeable, but it is still not objectionable for most applications.

Processing for effective speed gains greater than 2X will result in greater losses in maximum density (e.g., graying of the background in darkfield). In extreme cases, such as the photography of very faint fluorescence, these losses may be acceptable to gain information-recording ability.

KODAK Photomicrography Film (Process E-4)

KODAK Photomicrography Color Film 2843—Preferred for photographing very thin sections, weak stains, or where very high contrast is needed. (See Appendix IV for more details.) Special filtration required, typically—KODAK Color Compensating Filters CC40Y + CC40C + CC30C with tungsten-halogen microscope lamp (3200 K). Process E-4 can be done by the user, or exposed film can be sent to a commercial processor or a Kodak processing laboratory.

Photomicrography Color Film cannot be processed in Process E-6 chemicals.

KODACHROME Films

While KODACHROME Films can provide slightly differing rendition of certain stains, EKTACHROME Films provide equivalent resolving power and the bonus of user processing.

KODACHROME 25 Film (Daylight) and KODACHROME 64 Film (Daylight)—For general-purpose photomicrography on microscopes equipped with light sources providing daylight-quality illumination.

KODACHROME 40 Film 5070 (Type A)—Balanced for 3400 K photolamps. Requires use of light-balancing filter with typical microscope illumination.

KODACHROME Films are not user processable. Processing must be done by a commercial laboratory or by Kodak.

SPECIAL-PURPOSE FILMS FOR PHOTOMICROGRAPHY

KODAK Photomicrography Color Film 2483 (Process E-4)

This high-definition, high-contrast color reversal film is particularly useful to photomicrographers. The enhanced color saturation of both the reds and the blues will greatly improve the rendering of the most widely used histological stains. Chief among these are stains compounded of fuchsin and eosin. A didymium filter is not needed for best color reproduction. KODAK Photomicrography Color Film 2483 should be considered a supplement to, not a replacement for, other films for photomicrography.

HANDLING

135-36 Magazines—Load and unload the camera in subdued light. Rewind the film *completely* into the magazine before unloading.

Long Rolls—Handle only in total darkness. Because this film is thinner than conventional films, 125 feet of this film can be accommodated on the same size spool or in the same size chamber as would be occupied by 100 feet of conventional film.

Sheet film—Handle only in total darkness. Because this film is thinner than conventional sheet films, special precautions must be taken during exposure and processing. Before exposing, tap the lower edge of the film holder against a horizontal surface to settle the film against the bottom of the holder. This will prevent shifting of the film during exposure. For processing, use clip-type hangers; or when using channel-type hangers, tape the film to the hangers to prevent the film from becoming dislodged.

EXPOSURE

The exposure index is intended for use with through-the-lens meters with meter settings of the American National Standards Institute type. It includes the correction for the filter pack given. The exposure reading should be made without the filters. If it is necessary to make exposure readings *through* the filters, a somewhat higher exposure index may be used. This may vary depending on the equipment, but a meter setting of 12 is suggested as a start.

Exposure Latitude—As with other high-contrast films, exposure latitude is limited. For critical applications, bracket the final exposure with ± 1-stop adjustments, using half-stop increments.

PROCESSING

Recommended Process—Process E-4. *Total darkness* is required during early stages of processing. See instructions packaged with chemicals.

NOTE: Process E-3 may produce physical damage to the film and should not be used.

Laboratory Processing—Kodak offers a processing service for 135 magazines and 35 mm long rolls in Process E-4. Prepaid processing mailers are available from photo dealers. Film price does not include processing by Kodak, nor does Kodak authorize others to sell this film with any prepaid processing mailer attached.

Storage—Unexposed film should be stored at 13°C (55°F) or lower in the original sealed container. Aging effects are lessened by storing the film at lower temperatures. Film that must be kept for long periods of time should be stored at -18°C (0°F) or lower. Allow film to warm up to room temperature before opening the package. Process the film as soon as possible after exposure. Store processed film in a dust-free place at 21°C (70°F) or lower at a relative humidity of 30 to 50 percent. Protect film from strong light.

NOTE: Due to the birefringent properties of ESTAR Base, caution must be exercised when the processed film is used in an optical system where it will be located *between* polarizing elements. Most stereo projection systems are of this type. In such circumstances, the ESTAR Base may impart unwanted colored tints to the photograph.

Illumination	KODAK Color Compensating Filters	Exposure Time	Exposure Index
Tungsten-halogen microscope lamp (3200 K)	CC40Y + CC40C + CC30C	1/20 sec	5

Reciprocity Characteristics—

Exposure Time (Seconds)	3200 K tungsten Light Source					
	1/100	1/25	1/10	1	10	100
KODAK Color Compensating Filter	CC40Y + CC70C	CC40Y + CC70C	CC40Y + CC70C	CC50Y + CC70C	CC50Y + CC70C	CC50Y + CC70C
Exposure Time Increase	None	None	None	50%	150%	600%*

*Estimated

NOTE: These data should be used as a starting point only.

KODAK Instant Color Film/PR10

This is a daylight-balanced instant-print material that can be used with the KODAK Instant Film Back to produce finished color prints at the microscope. Since the film is balanced for daylight, exposure with 3200 K tungsten requires use of a KODAK WRATTEN Gelatin Filter No. 80A and an exposure increase of approximately 2 stops.

Longer exposures require adjustment for reciprocity effect. The approximate speed and color compensating filters are noted in the table. (Any adjustments are in addition to changes required to balance the light source.) For very long exposures, tests may be required.

Exposure time (sec)	1/100	1/10	1	10	100	1000
Exposure Index (EI)	160	125	80	25	10	5
Filter (CC)	None	5Y	20Y	30Y	30Y	30Y

KODAK Technical Pan Film 2415 (ESTAR-AH Base)

This black-and-white, panchromatic negative film with extended red sensitivity is useful in photomicrography. Contrast of this extremely fine grain, extremely high resolving power film can be varied with changes in development.

KODAK Technical Pan Film 2415 (ESTAR-AH Base) will be found useful in photomicrographic and photomacrographic situations in which conventional fine-grain films (such as KODAK PANATOMIC-X Film) have insufficient contrast. This film will provide several additional degrees of contrast for use with unstained specimens, phase contrast or other contrast-enhancing illumination, or at extreme magnifications. The film will also be useful in making black-and-white title slides, reduced copy negatives from black-and-white or color originals, and other applications where high resolution, high contrast, and maximum density are required. In photomicrography, a light-colored contrast filter such as KODAK WRATTEN Gelatin Filter No. 11 (yellowish-green) is suggested with most common histological stains in preference to the stronger filters often employed with other films (e.g., KODAK WRATTEN Gelatin Filter No. 58).

HANDLING

Load and unload the camera in subdued light.

Rewind the film *completely* into the magazine before unloading.

Darkroom Handling—Total darkness required. A KODAK Safelight Filter No. 3 (dark green) in a suitable lamp with a 15-watt bulb can be used for a few seconds only at 4 feet, after development is half completed.

EXPOSURE

The following exposure index (EI) values are intended as starting points for trial exposures to give satisfactory results with meters or photomicrography equipment having through-the-lens meters of the ANSI type. Bracketing exposures by half-stop intervals is suggested for first tests.

PROCESSING

Procedure for processing in small tanks with spiral reels using agitation at 30-second intervals:

1. Develop to the desired contrast index as specified in the section on "Exposure."

2. Rinse at 65 to 70°F (18.5 to 21°C) in KODAK Stop Bath SB-1a for 15 to 30 seconds.

3. Fix at 65 to 70°F (18.5 to 21°C), with frequent agitation.

> KODAK Rapid Fixer —1½ to 3 minutes
> KODAK Fixer —2 to 4 minutes
> KODAK Fixing Bath F-5—2 to 4 minutes

4. Wash in clear, running water at 65 to 70°F (18.5 to 21°C) for 5 to 15 minutes.

To save time and conserve water, use KODAK Hypo Clearing Agent. Rinse the fixed film in running water for 15 seconds. Next bathe the film in KODAK Hypo Clearing Agent for 30 seconds with agitation. Then wash the film for 1 minute in running water at 65 to 70°F (18.5 to 21°C), allowing at least one change of water during this time.

5. Dry in a dust-free place.

STORAGE

Store unexposed film at 70°F (21°C) or lower in the original sealed container. Aging effects are lessened by storing the film at lower temperatures. If film has been refrigerated or frozen, allow the package to reach room temperature for 2 to 3 hours before opening.

Store processed film in a cool, dry place.

Degree of Contrast Required	Contrast Index	KODAK Developer	Development Time at 68°F (20°C)	Exposure Index (Tungsten)
Maximum	2.40	D-19	4 minutes	125/22°
High	1.45	HC-110 Dilution D	8 minutes	100/21°
Moderate	0.85	HC-110, Dilution F*	8 minutes	50/18°

*HC-110 Developer, Dilution F, is prepared by diluting one part stock solution, mixed according to package instructions, with 19 parts water. Dilution F should be mixed fresh and discarded frequently, rather than replenished.

Reciprocity Characteristics—

Exposure Time (Seconds)	1/1000	1/100	1/10	1	10	100
Exposure Time Increase	None	None	None	None	60%*	150%*

*Slight contrast reduction usually does not require development increase.

NOTE: These data should be used as a starting point only.

Filter Factors—When a filter is used, determine the normal exposure without the filter. Then multiply the normal exposure by the filter factor given below.

KODAK WRATTEN Gelatin Filter	Color of Filter	Filter Factor
No. 11	Yellowish-Green	5
No. 12	Deep Yellow	1.25
No. 13	Dark Yellowish-Green	6.4
No. 25	Red	2
No. 47	Blue	25
No. 58	Green	12.5

INDEX